International Conference for Free and Open Source Software for Geospatial (FOSS4G 2017)

OSGeo Journal Volume 17, Issue 1

Boston, Massachusetts, USA
14-19 August 2017

Editors:

Charles M. Schweik
Mohammed Zia
Andy Anderson

ISBN: 978-1-7138-1903-5

Printed from e-media with permission by:

Curran Associates, Inc.
57 Morehouse Lane
Red Hook, NY 12571

CURRAN ASSOCIATES INC. proceedings.com

Some format issues inherent in the e-media version may also appear in this print version.

OSGeo
9450 SW Gemini Dr. #42523
Beaverton, Oregon
United States
97008

www.osgeo.org

Additional copies of this publication are available from:

Curran Associates, Inc.
57 Morehouse Lane
Red Hook, NY 12571 USA
Phone: 845-758-0400
Fax: 845-758-2633
Email: curran@proceedings.com
Web: www.proceedings.com

Free and Open Source Software for Geospatial (FOSS4G) Conference Proceedings

Volume 17 *Boston, USA* Article 2

2017

FOSS4G 2017 Full Conference Proceedings (papers and posters)

Charles M. Schweik
University of Massachusetts - Amherst

Mohammed Zia
National Innovation & Research Center for Geographical Information Technologies, Turkey

Andy Anderson
Amherst College

FOSS4G 2017 Academic Track Proceedings co-editors:

Charles M. Schweik (University of Massachusetts, Amherst, USA)

Mohammed Zia (National Innovation & Research Center for Geographical Information Technologies, Turkey)

Andy Anderson (Amherst College, Massachusetts, USA)

FOSS4G 2017 Academic Track Reviewers:

Academic Program

Papers and Posters

August 14-19, 2017

OSGeo Journal Volume 17, Issue 1

The online version is published at http://scholarworks.umass.edu/foss4g/vol17/iss1 and as Volume 17, Issue 1 of the OSGeo Journal at https://journal.osgeo.org/index.php/journal.

Editors: Charles M. Schweik (University of Massachusetts, Amherst, USA)
Mohammed Zia (National Innovation & Research Center for GIS, Turkey)
Andy Anderson (Amherst College, Massachusetts, USA)

From the Editors...

Greetings! This Conference Proceedings is a collection of outstanding papers and posters submitted to the Academic Program of the International Conference for Free and Open Source Software for Geospatial (FOSS4G), 14th to 19th August 2017 in Boston, U.S.A.[1]

As Barend Köbben noted in his opening of the 2014 Proceedings[2], FOSS4G has not, historically, been identified as an academic conference. It was founded by FOSS4G developers affiliated with the Open Source Geospatial Foundation (OSGeo), with a primary focus on sharing new developments and supporting conversations between developers and their user communities. But a quick review of FOSS4G conference proceedings demonstrates that the academic side of FOSS4G is growing, which is not surprising given how FOSS4G has matured over the last decade. In 2011 at FOSS4G Denver, seven academic papers were published in the academic proceedings[3]. Four years later, at 2015 FOSS4G Seoul, the academic proceedings contained eleven papers[4]. This year, FOSS4G 2017 Boston had 49 academic abstracts submitted amongst the 416 all-conference submissions. After conducting a peer-review process[5], the Academic Program Committee selected eighteen papers and nine posters for publication in these proceedings[6]. Many will agree that this is an important trend, for in academics and in geospatial science more generally, the phenomena of FOSS4G needs to be strongly represented to increase their accessibility and expand their audiences.

Substantively, readers will discover that these academic papers cover a wide range of topics. For example, under the general topic of geospatial modeling, several focus on hydrologic modeling (papers 1.7 and 1.11 in the table of contents), one on multi-scale urban climate modeling (paper 1.18), and one modeling the traveling salesman problem using OpenStreetMap data (paper 1.12). Research related to spatial databases are represented, with one comparing the performance of SQL versus NoSQL spatial databases (paper 1.1), one on the optimization of spatiotemporal indexing (paper 1.10), and one on the storage and management of digital maps via an open geoportal library system (paper 1.9). Other academics or geospatial scientists submitted papers that investigate approaches to geovisualization, including one on the processing and visual display of big geospatial data (paper 1.13) and one that develops methods to produce beautiful geospatial data in map output that mimic the brush strokes of painter Vincent Van Gogh (paper 1.17). Still other papers move forward geospatial research methods (papers 1.6 and 1.16), or apply FOSS4G technologies and open standards in environmental science applications (papers 1.4 and 1.5). In all cases, FOSS4G developmental approaches and technologies are well represented.

In addition to creating a new FOSS4G Proceedings that will follow tradition and be published on the OSGeo Journal website[7], we also have implemented, with the help of the University of Massachusetts Amherst Scholarly Communication team[8], a new paper review and proceedings publishing system built upon BEPress' Digital Commons system[9]. The UMass Amherst ScholarWorks platform provides a behind-the-scenes paper and poster review processing system that allows Proceeding Editors to manage initial submissions, assign reviewers, process

[1] http://2017.foss4g.org/

[2] OSGeo Journal Volume 14, available at https://journal.osgeo.org/index.php/journal/issue/viewIssue/35/6

[3] https://journal.osgeo.org/index.php/journal/issue/view/33/showToc

[4] http://www.meci.co.kr/societyevent/FOSS4G2015/download/FOSS4G%20PROCEEDINGS%20ver2.pdf

[5] http://2017.foss4g.org/program/academic-review/

[6] The papers are published in this volume. The posters are available in the online edition, at http://scholarworks.umass.edu/foss4g/vol17/iss1

[7] https://journal.osgeo.org/index.php/journal

[8] https://www.library.umass.edu/locations/sc/

[9] http://scholarworks.umass.edu/foss4g

reviews, communicate to authors, and, ultimately, produce the online proceedings that can be made available both through the OSGeo Journal[10] and through this new FOSS4G Proceedings portal[9]. The benefit of the latter is that the proceedings metadata are automatically entered into a wide variety of library databases, making them much more easily discovered by people using library database search systems. Importantly, we have developed this system not only to support our work for this Proceedings, but with the hope that academic track editors of future FOSS4G regional and global conferences will utilize this free - as in cost - system made available by the University of Massachusetts Amherst.

We need to conclude this introduction by extending our sincere thanks to the paper and poster authors, and the extremely responsive Academic Committee[11] who took their time to provide the authors with reactions and suggested improvements to draft paper submissions. We also need to thank FOSS4G conference Operations Coordinator Ashley Tardiff for her help in the review process, and to Aurora Arder of bepress and University of Massachusetts Librarian Erin Jerome for their help in setting up the review and publishing Scholarworks infrastructure[9], and to the University of Massachusetts Amherst Libraries for provision and support of this system, now and into the future.

We hope readers enjoy these proceedings, and are inspired to continue to build and grow the academic program generally, and work to develop standard academic sub-tracks such as Geospatial Modeling, Spatial Database research, Geovisualization, and Environmental Science Applications in FOSS4G conferences in the future. We also hope that future editors of regional and global FOSS4G conferences will take advantage of the review and publishing system we have introduced[9]. Future editors should contact editor Charlie Schweik[12] for instructions and access.

Sincerely,

Charlie Schweik
Mohammed Zia
Andy Anderson
FOSS4G Academic Committee Co-chairs and Proceedings Co-editors

[10]https://journal.osgeo.org/index.php/journal
[11]https://wiki.osgeo.org/wiki/FOSS4G_2017#Academic_Committee
[12]cschweik AT pubpol DOT umass DOT edu

Contents

Papers

Analyzing the performance of NoSQL vs. SQL databases for Spatial and Aggregate queries

Sarthak Agarwal[a,*], KS Rajan[a]

[a]*International Institute of Information Technology Hyderabad Gachibowli, Hyderabad, India*

Abstract: Relational databases have been around for a long time and spatial databases have exploited this feature for close to two decades. The recent past has seen the development of NoSQL non-relational databases, which are now being adopted for spatial object storage and handling, too. While SQL databases face scalability and agility challenges and fail to take the advantage of the cheap memory and processing power available these days, NoSQL databases can handle the rise in the data storage and frequency at which it is accessed and processed - which are essential features needed in geospatial scenarios, which do not deal with a fixed schema(geometry) and fixed data size. This paper attempts to evaluate the performance of an existing NoSQL database 'MongoDB' with its inbuilt spatial functions with that of a SQL database with spatial extension 'PostGIS' for two problems spatial and aggregate queries, across a range of datasets, with varying features counts. All the data in the analysis was processed In-memory and no secondary memory was used. Initial results suggest that MongoDB performs better by an average factor of 10x-25x which increases exponentially as the data size increases in both indexed and non-indexed operations. Given these results, NoSQL databases may be better suited for simultaneous multiple-user query systems including Web-GIS and mobile-GIS. Further studies are required to understand the full potential of NoSQL databases across various geometries and spatial query types.

*Corresponding author
Email address: `sarthak.a@research.iiit.ac.in` (Sarthak Agarwal)

1. Introduction

Traditionally Databases were designed to structure and organize any form of data. However, as the database size increased and to optimize databases for the geospatial domain we use spatial databases. Satellite images are one prominent example of spatial databases Govind and Sharma 2013. To extract spatial information from a satellite image, it has to be processed in a spatial frame of reference. However, satellites are not the only type of spatial databases. Maps are also stored in the spatial database.

Like other database systems, spatial databases have also relied on the relational databases to handle and manage spatial objects and their associated attribute information. They have been of great value in cases where we have a defined structure of our schema, including geometrical characteristics of the spatial objects. In addition, in many spatial applications, we do not always have a fixed schema, there can be many geometries with different shape and the requirements evolve as the data size increases depending on the case scenario where relational Databases can limit the potential use or design of the solution.

For the satisfaction of the user's significant characteristics of a database such as scalability, performance and latency play a crucial role. Especially social media projects, such as Facebook and Google+, with high user traffic, use other database management systems. such as Apache Cassandra or Google BigTable. Instead of the relational approach, a Not-only-SQL (NoSQL) method is used. NoSQL-databases are increasingly used to deal with simultaneously high read and write requests related to large datasets.

While SQL databases face scalability and agility challenges and fail to take the advantage of the cheap memory and processing power available these days, NoSQL databases can handle the rise in the data storage and frequency at which it is accessed and processed - which are essential features needed in geospatial scenarios, which do not deal with a fixed schema(geometry) and fixed data size Loureno et al. 2015.

NoSQL data stores may provide advantages over relational databases, However, they generally lack the robustness of relational databases for those advantages. The aim of this paper is to discover some benefits of a selected NoSQL data store as compared to a traditional relational database when storing and querying spatial vector data. Our work is influenced by the work done previously on the similar theories de Souza Baptista et al. 2011, Xiao and Liu 2011, Schmid et al. 2015 and some practical implementation of those theories Popescu and Bacalu 2009, Steiniger and AndrewJ.S.Hunter 2012, van der Veen et al. 2012.

One of the very important problems we face is of aggregate queries i.e when we combine the normal non-spatial data with the spatial data. Here we are trying to solve a similar problem based on our previous studies. The aim of this problem is to combine the real time spatial and non-spatial scenarios into one and then analyze the performance of SQL vs NoSQL databases in such scenarios.

The two problems are carefully chosen which are, first the total number of restaurants in an area which is vegetarian only. This problem has two very important features. Firstly we have to use containment query to find out the total number of restaurants in an area and apply sum aggregate query on such restaurants which are vegetarian. The second problem is getting the total number of distinct cuisine in an area. This problem also has very important aggregate feature i.e. distinct in a column. So using these queries we are able to simulate most of the real case scenarios and analyzing the performance of each database.

2. NoSQL vs SQL in spatial context

2.1. NoSQL vs SQL

SQL Server and relational databases have been the go-to databases for over 20 years. The increased need to process higher volumes, velocities, and varieties of data at a rapid rate has altered the nature of data storage needs for application developers. SQL Server and relational databases were the most popular databases. However, in order to enable the present day processing needs, NoSQL databases have gained popularity due to their ability to store unstructured and heterogeneous data at scale. Relational databases still remain a popular default option due to their easy to understand table structure, however there are many reasons to explore beyond relational databases.

NoSQL is a category of databases distinctly different from SQL databases. NoSQL is often used to refer to data management systems that are Not SQL or an approach to data management that includes Not only SQL". There are a number of technologies in the NoSQL category, including document databases, key value stores, column family stores, and graph databases, which are popular with gaming, social, and IoT apps.

2.2. NoSQL in spatial context

If we talk in the context of spatial Informatics, we need databases at multiple instances. We need databases to store maps, in maps particularly the schema can be vague. We also need the databases to store images from satellites and another kind of spatial data transmitted by a satellite such as weather info. So in short spatial informatics needs to store a lot of data. So to solve this problem we need some kind of data store.

2.3. Current systems, A case study

There are currently many NoSQL databases systems available but only few support spatial databases. In this section, we are going to discuss some of those systems and what all function so each of them offers.

MongoDB uses currently two geospatial indexes, 2d, and 2dsphere. The 2d index is used to calculate distances on a plane surface. The 2dsphere index calculates geometries over an Earth-like sphere. The coordinate reference system is currently limited to the WGS84 datum. MongoDB computes the geohash values for the coordinate pairs and then indexes the geohash values. A precise description of the indexing techniques of the geohash values is not available at the moment Anonymous 2011-2015.

CouchBase supports indexing of two-dimensional data using an R-Tree index. CouchBase, therefore, provides spatial views which enable a geospatial query using bounding boxes Ostrovsky and Rodenski 2014. A precise description of the indexing techniques in CouchBase is currently also not available.

Besides the storage and indexing of spatial data, the query process is an important aspect. For querying spatial data several geo-functions are available in relational databases. They enable different queries with geo-context at the database level using SQL. An example for a geo-function is the calculation of a buffer around a point feature or a line feature.

For the relational-database PostgreSQL, there is a special extension available, PostGIS, for integrating several geo- functions. MongoDB and CouchBase don't have a separate extension at the moment but they support some geo-functions. Table 1 compares the geo-functions of the three databases.

PostgreSQL/PostGIS inherits more than one thousand geo- functions. Table 1 includes only a selection of them. MongoDB only supports three geo-functions, $geoWithin, $geoIntersects and $near. The MongoDB $geoWithin operator corresponds to the ST_Within function in PostgreSQL/PostGIS, and the MongoDB $geoIntersects operator corresponds to the function ST_Intersects in PostgreSQL/PostGIS.

The function $near delivers the next located geometry for a predefined point. The $near function can be used in combination with a $maxDistance parameter. In that case, MongoDB delivers all geometries within a certain distance ordered by the distance. PostgreSQL can calculate this using the ST_DWithin function. The results however, need to be additionally ordered by the distance.

CouchBase can only query point geometries within a BoundingBox (BBox). The BBox-function of CouchBase can be compared to the $geoWithin (MongoDB) and ST_Within (PostGIS) functions, however, MongoDB and PostGIS can use different polygons, not only an axial parallel polygon.

The overview of the implemented geo-functions shows that PostgreSQL with its extension PostGIS has the most comprehensive geo-functionalities with more than one thousand functions. For a complete list of all implemented functions, it is referred to the PostGIS handbook. The two NoSQL databases have very limited implemented geo-functions. MongoDB just implements three functions whereas CouchBase just implements one geo-function.

2.4. Scope of improvements

The NoSQL spatial database is still at their inception. There is a lot of research and development yet to be done in both theoretical part and practical part of them. Currently, they support only a limited type of geometry and very fewer functions to manipulate them. We have to build the design of these databases in such a way that adding more functions can be flexible. We can take inspiration from the functions currently present in PostGIS and can start implementing those functions in MongoDB as well. spatial databases also currently support for various indexes, Xiang et al. 2016 illustrates the implementation of R-trees in MongoDB for spatial indexing.

PostGIS	**MongoDB**	**CouchBase**
ST_Within	$geoWithin	BBOX
ST_Intersects	$geoIntersects	
ST_DWithin + Order by dist	$near + param(Distance)	
ST_Area		
....		

Table 1: Geo-Functions of the investigated databases.

3. Dataset

For this analysis, the dataset was custom generated.

3.1. All vegetarian restaurants in an area

Point Containment problem works for any geometry and reports whether the given geometry is completely inside another geometry or not. This is a vital and traditional problem on spatial databases. It is useful in the domain of map generations, modeling, analyzing spatial data over an area, for example, we want to report how the number of houses has changed over time in a city by analyzing the spatial data of that town. We can count the number of points (which represents each house in this case) in the polygon(city) for all the years using point within a polygon problem.

The dataset D1 consisted of two layers, the first layer simulating the restaurants which are represented as points having some non-spatial attributes such as timing, price, cuisine, etc. Another layer consisted of square boxes of different perimeter scattered both sequentially and randomly over the space with few points completely inside the box and others completely outside the box simulating the area in interest.

3.2. Distinct Cuisine in an area

Like the previous problem here also we use Point Containment query which works for any geometry and reports whether the given geometry is completely inside another geometry or not.

The dataset D2 consists of similar layers one simulating restaurants and other simulating the area in interest.

4. Performance

Tests were run on both of the database systems with same datasets one with Index and other without an index. The time recorded is in seconds for both indexed and non-indexed analysis. Lines and polygons in PostGIS were indexed using GIST index method and in MongoDB they were indexed using 2d Sphere indexing method.

Test Setup

All the data in the analysis was processed In-memory and no secondary memory was used.

Hardware used-

- Ram 16 Gb
- Processor - Intel Core i7-5500u CPU @ 2.40 Ghz x 4
- OS Ubuntu 14.04 64 bit
- Disk Solid state hard drive

Softwares used-

- PostgreSQL 9.3.12
- PostGIS 2.1
- MongoDB 3.2.5

For Post results analysis Libre Office and Google sheets were used to plot graphs.

If we observe the graph from figure 1 we will notice that MongoDB works much faster than PostGIS in all the cases(from small datasets too large datasets). The difference in performance is in the order of 10(approx.). Also, another observation we make from the graph is that the performance difference between indexed and non-indexed queries is small in MongoDB as compared to PostGIS. So we reach a conclusion that the performance of both indexed and non-indexed Line Intersection queries is better in the case of MongoDB as compared to PostGIS.

In experiment 2, firstly the performance was analyzed without indexing any geometry and time was observed which suggests that MongoDB performs better as the data size increases whereas PostGIS does not perform as well with huge datasets and time complexity increases exponentially. However after indexing the geometries performance of both the database engine improved by a substantial factor.

If we observe the results of the experiments, we will notice that indexed datasets perform better than non-indexed dataset and PostGIS time increases the exponentially as the size of dataset increases whereas MongoDB still performs within some bounds.

5. Discussions and Conclusions

The experiment analyses the performance difference and ease of implementation while developing real time application involving both spatial and non-spatial use cases. These results suggest that MongoDB performs better by an average factor of 10 which increases exponentially as the data size increases in both indexed and non-indexed operations for problems. Given these results, NoSQL databases may be better stated for simultaneous multiple-user query systems including Web-GIS and mobile-GIS.

This implies that non-relational databases are more suited to the multi-user query systems and has the potential to be implemented in servers with limited computational power. Further studies are required to identify its appropriateness and incorporate a range of spatial algorithms within non-relational databases.

The focus of our research is to mainly benchmark the real case scenarios including both spatial and non-spatial use cases. The motivation behind this work is to minimize the dependency for server during mobile routing. We want the client itself to act as a server during routing and for that we need a light database system that can be easily ported to the mobile devices, so here we are comparing the performance of SQL vs. NoSQL databases which are comparatively lighter.

However there are still some limitations on using NoSQL databases over SQL databases. There are not as many spatial functions in NoSQL as in SQL. The currently implemented geo-functions support only very basic operations. Relational databases are still far superior if the user needs to calculate geoinformation on database level Schmid et al. 2015. The results presented in the paper are only valid for the chosen database settings but they clearly show that NoSQL databases are a possible alternative, at least for querying attribute information. PostgreSQL also has an implementation of NoSQL of its own, however it does not support PostGIS currently however we can export the results of PostGIS queries as GeoJSON objects. In future, we are planning on expanding our study to other spatial query functions as well as spatial algorithms such as shortest path problem to evaluate the performance of NoSQL on such platforms and also test the performance of MongoDB on distributed systems.

References

Anonymous, 2011-2015. Mongodb, inc.; geospatial indexes and queries. Online, online; accessed July 21, 2017. URL `http://docs.mongodb.org/manual/applications/geospatialindexes/`

de Souza Baptista, C., de Oliveira, M. G., da Silva, T. E., 2011. Using ogc services to interoperate spatial data stored in sql and nosql databases. XII GEOINFO, Campos do Jordo, Brazil.

Govind, S., Sharma, A., 2013. Open source spatial database for mobile devices. Computer Engineering and Intelligent Systems 4(6).

Loureno, J. R., Cabral, B., Carreiro, P., Vieira, M., Bernardino, J., 2015. Choosing the right nosql database for the job: a quality attribute evaluation. Journal of Big Data 2:18.

Ostrovsky, D., Rodenski, Y., 2014. Pro couchbase server. Apress.

Popescu, A., Bacalu, A.-M., 2009. Geo nosql: Couchdb, mongodb, and tokyo cabinet. Online, online; accessed July 21, 2017.
URL `http://nosql.mypopescu.com/post/300199706/geo-nosql-couchdb-mongodb-tokyo-cabinet`

Schmid, S., Galicz, E., Reinhardt, W., 2015. Performance investigation of selected sql and nosql databases. AGILE 2015, Lisbon.

Steiniger, S., AndrewJ.S.Hunter, 2012. Free and open source gis software for building a spatial data infrastructure. Geospatial Free and Open Source Software in the 21st Century(Part 5).

van der Veen, J. S., van der Waaij, B., Meijer, R. J., 2012. Sensor data storage performance: Sql or nosql, physical or virtual. IEEE Fifth International Conference on Cloud Computing, Honolulu, Hawaii, USA.

Xiang, L., Shao, X., Wang, D., 2016. Providing r-tree support for mongodb. The International Archives of the Photogrammetry, Remote Sensing and Spatial Information Sciences, Prague, Czech Republic XLI-B4.

Xiao, Z., Liu, Y., 2011. Remote sensing image database based on nosql database. Geoinformatics.

Experiment 1. Performance of All vegetarian restaurants in an area

Restaurants in 1st Layer	Boxes in 2nd Layer	No. of vegan restaurants within	PostGIS time(s)	PostGIS Index time(s)	MongoDB time(s)	MongoDB Index time(s)
10	5	5	9.25	8.39	6	7
10	10	17	17.63	14.82	12	12
10	20	17	17.14	14.76	27	26
10	40	17	15.01	17.55	52	55
20	5	5	23.18	21.86	7	9
20	10	17	51.46	47.88	13	14
20	20	45	108.7	133.71	27	28
20	40	46	147.07	151.24	50	48
40	5	5	72.21	78.02	8	6
40	10	5	146.59	183.91	12	13
40	20	45	727.36	630.53	28	27
40	40	122	1092.52	1279.14	51	52
40	60	122	1272.77	1065.34	96	97
60	5	5	120.92	121.71	6	6
60	10	17	292.7	313.96	13	11
60	20	45	1130.5	1102.48	28	30
60	40	122	3111.69	2920.86	51	52
60	60	204	3753.86	3427.31	98	102
60	80	204	3069.14	3213.61	99	102
80	40	122	4573.4	3269.07	53	53
80	60	204	5325.57	2471.84	102	100
80	80	298	NC*	NC*	106	109
80	90	298	NC*	NC*	144	123

Table 2: Average Time for Vegan in an area problem (*NC- Not Computable).

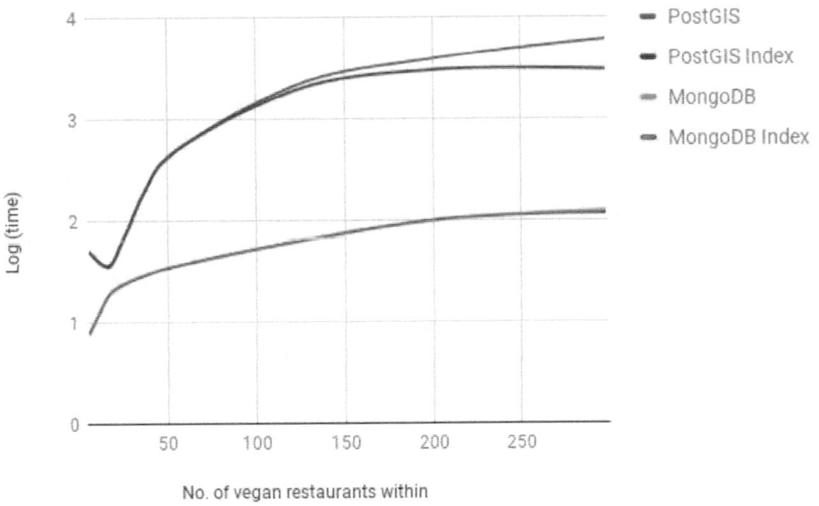

Figure 1: Graph of No. of restaurants vs average time for all datasets in D1.

Experiment 2. Performance of Distinct Cuisine restaurants in an area

Restaurants in 1st Layer	Boxes in 2nd Layer	No. of Distinct Cuisine restaurants within	PostGIS time(s)	PostGIS Index time(s)	MongoDB time(s)	MongoDB Index time(s)
10	5	5	8.96	8.72	5.89	7.26
10	10	17	18.7	13.98	12.59	12.05
10	20	17	16.95	14.36	27.77	28.27
10	40	17	14.87	17.8	55.8	59.5
20	5	5	24.62	20.31	6.52	8.29
20	10	17	46.6	49.34	11.98	15.35
20	20	45	104.67	142.97	28.15	28.28
20	40	46	142.55	142.75	54.09	50.08
40	5	5	75.45	75.49	8.01	6.19
40	10	5	136.68	190.67	11.91	11.99
40	20	45	704.79	584.84	30.17	27.82
40	40	122	1027.05	1361.62	49.23	55.49
40	60	122	1379.45	1075.67	94.82	104.5
60	5	5	114.92	126.7	5.66	6.26
60	10	17	301.36	337.02	11.94	10.67
60	20	45	1091.33	1119.77	26	29.18
60	40	122	3275.41	2746.24	49.6	50.24
60	60	204	3721.13	3302.4	98.38	98.32
60	80	204	3052.97	3186.87	103.85	104.62
80	40	122	4775.85	3169.52	57.3	49.25
80	60	204	4827.24	2249.57	111.63	105.23
80	80	298	6384.51	2276.98	113.78	114.94
80	90	298	6100.11	2518.85	134.74	122.6

Table 3: Average Time for Distinct Cuisine problem.

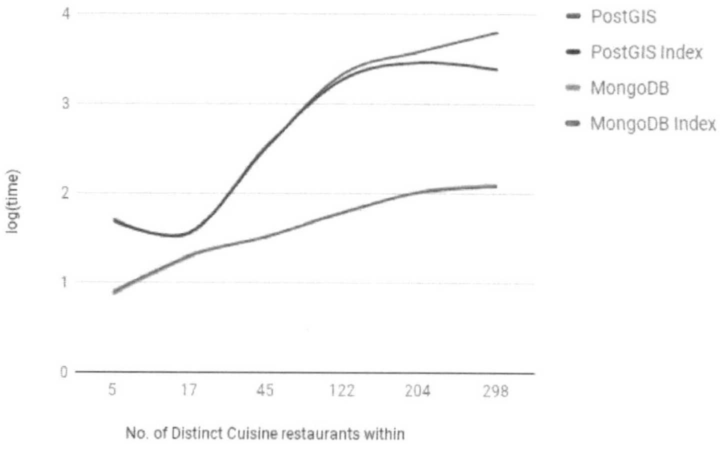

Figure 2: Graph of No. of restaurants vs average time for all datasets in D2.

A New Spatial Approach for Efficient Transformation of Equality - Generalized TSP to TSP

Mohammed Zia[a,c,*], Ziyadin Cakir[b,c], Dursun Zafer Seker[a,c]

[a] *Geomatics Engineering Department, Faculty of Civil Engineering*

[b] *Department of Geology, Faculty of Mines*

[c] *Istanbul Technical University, Turkey*

Abstract: The Equality - Generalized Travelling Salesman Problem (E-GTSP) asks to find a *Hamiltonian* cycle visiting each group exactly once, where each group represents a type of visiting node. This can represent a range of combinatorial optimization problem of NP-hard type like planning, logistics, etc. Its solution requires transformation of E-GTSP to TSP before solving it using a given TSP solver. This paper presents 5 different search-algorithms for optimal transformation which considers spatial spread of nodes of each group. Algorithms are tested over 15 cities with different street-network's fractal-dimension for 5 instances of group-counts each. It's observed that the R-Search algorithm, which selects nodes from each group depending upon their radial separation with respect to the start-end point, is the optimal search criterion among all other algorithms with a mean length error of 8.8%. This study will help developers and researchers to answer complex routing problems from a spatial perspective.

*Corresponding author
Email address: zia@itu.edu.tr,mohammed.zia33@gmail.com (Mohammed Zia)
URL: https://linkedin.com/in/zia33 (Mohammed Zia)

1. Introduction

The Travelling Salesman Problem (TSP) is one of the well-known and extensively studies combinatorial optimization problem which demands to estimate the shortest path in terms of length, time, etc. that visits each station/node/vertex in a given group/set exactly once before returning to the starting point. Mathematically, it could be stated as follows: Given a directed/undirected graph $G = (V, E)$ with set of vertices V and set of weighted edges E, find the shortest path between start vertex s and end vertex e (e could be same as s for closed path) that visits each vertex for a given set $V' \subseteq V - \{s, e\}$ exactly once. It is nothing but calculating the minimum-cost *Hamiltonian* cycle in G Hamilton 1856. It has always remained a great source of attraction due to its ranging areas of application like routing, communication networking, sequencing and scheduling, etc. Lenstra and Kan 1975. A detailed classification of types of TSP and possible solutions can be found at Psaraftis et al. 2016.

A variety of heuristic and tabu search algorithms which are constructive are developed by researchers in the past to solve this NP-hard problem, like Nearest Neighbour Gutin and Yeo 2007, Clarke-Wright heuristic Clarke and Wright 1964, Minimum Spanning Tree heuristic (eg. Kruskal's algorithm) Rosenkrantz et al. 2009, Christofides heuristic T. and Roberto 2015. Other algorithms include • K-OPT Croes 1958, • Tabu search Glover 1990, • Simulated Annealing Kirkpatrick et al. 1983, • Held-Karp lower bound Valenzuela and Jones 1997, • Lin-Kernighan algorithm Helsgaun 2000, • Lin-Kernighan-Helsgaun Helsgaun 2009, and • Cutting Plane and Branch-Bound techniques. A good description is also given in TSP Wiki page and Curtin et al. 2014.

Let $G = (V, E)$ be a graph where $V = \{v_1, v_2, ..., v_n\}$ is the set of vertex, $E = \{(v_i, v_j) \mid i \neq j; v_i, v_j \in V\}$ is the edge set, and $W = \{w_{ij}\}$ is the non-negative cost or weightage defined on E. If E is undirected then directions are irrelevant and $(v_i, v_j) = (v_j, v_i)$. Now, in the GTSP, V is partitioned into x mutually exclusive and exhaustive groups such that $V^g = \{V_1, V_2, ..., V_x\}$ and $V = V_1 \cup V_2 \cup ... \cup V_x$ with $V_\alpha \cap V_\beta = \varnothing$ for all $\alpha, \beta = 1, 2, ..., x$ and $\alpha \neq \beta$. The Generalized-TSP (GTSP) asks to determine the shortest *Hamiltonian* circuit passing through each group *at least once* (introduced independently by Henry-Labor 1969, Saksena 1970, and Srivastava et al. 1969) or *exactly once* (by Noon and Bean 1991, and Laporte and Nobert 1983). The *exactly once* case of GTSP is known as the Equality-GTSP or E-GTSP Helsgaun 2015, where the route contains exactly one station from each V^g group. It has prime relevance in location-based problems, urban planning, postal routing, logistics, microchips manufacturing, telecommunication problems, and railway track optimization. Its existing solutions, however, are difficult to replicate by general users for TSP models representing GIS problems, primarily vehicle-routing.

In order to transform the E-GTSP to TSP, authors have tried to think from a spatial perspective as the underlying model primarily represents an urban setup. Once done, it could be solved using the state-of-the-art Lin-Kernighan-Helsgaun TSP solver Helsgaun 2015. Five different search algorithms are presented and analysed on OpenStreetMap (OSM) street dataset for best possible transformation. Different group-counts (i.e. $|V^g| = 1, 2, 3, 4, 5$) are modelled to test proposed algorithms at different complexity levels and results are discussed in the Results and Discussion Section 5. In the following sections, the proposed search algorithms are explained and tested, and finally a commentary is provided in the Conclusion Section 6.

2. Transformation of E-GTSP to TSP

The GTSP was introduced through the record balancing problems aroused in computer design by Henry-Labor 1969, Srivastava et al. 1969, and Saksena 1970. The shortcoming of E-GTSP to TSP transformation is that it increases problem's dimension drastically, and therefore it becomes practically unfeasible to solve. Dynamic programmatically, an E-GTSP's solution is based upon deciding the following two **decisions** in written order:

1 Selection of a vertex subset V^s, also termed as station, such that $V^s \subseteq V$ and $V^s \cap V_\alpha = 1$ for all $\alpha = 1, 2, ..., x$. *Note* that, $|V^s| = |V^g|$.

2 Calculation of the minimum-cost *Hamiltonian* cycle in subgraph $G^s = (V^s, E^s)$ of G produced by V^s.

The presented search algorithms do not increase problem's size by increasing vertex or edge count. It is much easier to filter-out distant and sub-optimal vertices from each group considering their spatial spread with respect to the end vertices. In this article, *points* represent all possible vertices of V including start/end, *group* represents each set of vertex from V exhibiting one particular type of vertex, and *station* represents the selected point from each group.

3. Methodology

Five different search criteria to select one point from each group are presented here (Fig. 1). Although a number of such algorithms are possible, authors believe presented ones to be mutually exclusive and covering a range of search possibilities.

3.1. E-Search and D-Search

Euclidean-Search (E-Search) is based upon the euclidean line between start-end points in a given instance. The basic approach behind this is to connect both the start/end points by a straight line and select the closest point from each group with respect to this line (Fig. 1). It involves two stages before reducing E-GTSP to TSP. It is $O(n)$, where $n = |V|$.

Dijkstra-Search (D-Search), on the other hand, involves calculating Dijkstra route between given start/end points before selecting the closest point with respect to that route (Fig. 1). Computationally it is more expensive than E-Search, with $O(n^2)$. Algorithm 1 is the pseudocode of the above two search algorithms.

3.2. R-Search

Radial-Search (R-Search) is based upon the radial distance of stations from start/end points for a given E-GTSP instance. Closest point from each group is selected twice, making $|V^s| = 2 \times x$ (Section 2), leading 2^x different TSPs, making **decision 2** computationally expensive (Fig. 1). It involves an O-complexity of $O(n)$. Furthermore, it evaluates points lying outside the proximity of start-end region more efficiently, unlike E-Search and D-Search.

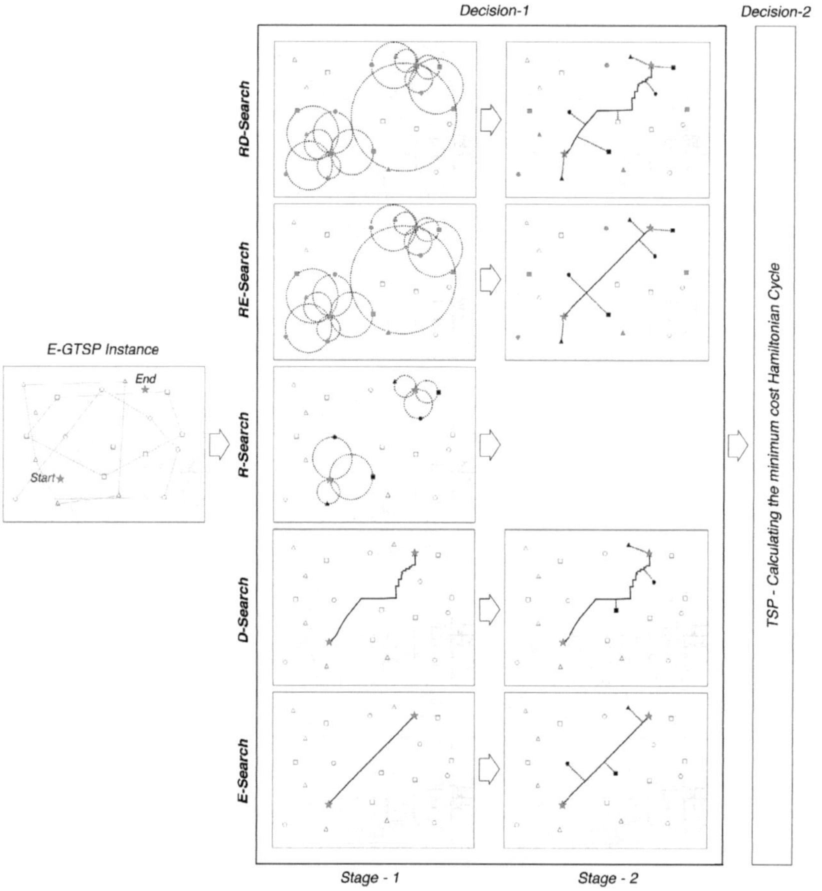

Figure 1: Diagrammatic representation of 5 different proposed search algorithms. Three groups of different kind of points, marked by oval, triangular, and rectangular markers, are used to develop an E-GTSP instance. It should be noted that for *R-Search*, *RE-Search*, and *RD-Search* there are two stations from each group after **decision 1** (Section 2), thus, generating 8 ($2 \times 2 \times 2$, for drawn instance) possible TSPs for **decision 2**.

3.3. RE-Search and RD-Search

Finally, the last two search algorithms are a hybrid of R-Search and E-Search (RE-Search), and R-Search and D-Search (RD-Search). They first find the radially closest two points from each group with respect to both the start/end points independently, making $|V^s| = 2 \times 2 \times x$. Later they select the closest stations from both the ends with respect to the euclidean line (for RE-Search) or Dijskstra route (for RD-Search), Fig. 1. They have the complexities of $O(n)$ and $O(n^2)$, respectively. Similar to R-Search, the second step leads to a total of 2^x TSPs. Once estimated, the optimal V^s from each group it is solved by Simulated Annealing Kirkpatrick et al. 1983 TSP Solver.

4. Study Area and Data-Set Used

OSM dataset of 15 cities belonging to five different fractal dimension (frac-D) bins were used to test proposed algorithms, out of 210 cities worldwide. Frac-D value is directly proportional to road density of a region with 1-D representing area with just one road section and 2-D representing area completely filled up with roads. Bin size is kept 0.1 for highly resolved classes

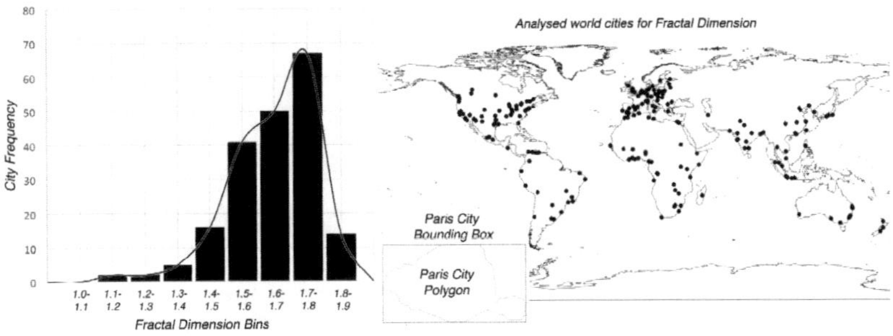

Figure 2: Histogram showing the frequency of cities (marked on the world map *right*) for each Fractal-Dimension bin. Increase in dimension represents increase in road-density. 3 cities are selected from each top 5 bins to test the proposed search algorithms. Each city-polygon's bounding box is used to download corresponding OSM XML data from its Overpass API.

(Fig. 2). **F**ractal **D**imension **C**alculator is used to calculate the dimension of each city Bourke 2003. All 105 start-end points to test algorithms are randomly generated and 5 points from each group are likewise selected from OSM point dataset to better represent physical stops in the city. Observations regarding best search algorithm with increased complexity are discussed in Section 5.

5. Results and Discussion

5 different instances, i.e. $|V^s|$, are modelled for each city to test each algorithm. Fig. 3a is a 3D-plot between *different search algorithms* used, *different number of stations to be visited*, i.e. $|V^g|$, and *different type of street-networks*, where each circle represents the average fractional error in E-GTSP route-length coming out of all 105 start-end pairs estimated with respect to the optimal route (by brute-force). There are 375 (`15cities` $\times$ `5group-counts` $\times$ `5algorithms`) colored circles, with black ones representing error more than 20%. It is clear from the horizontal color tone variation that the percentage error in route length will increase with higher $|V^s|$ instances (marked by white arrow Fig. 3a) irrespective of the algorithm used. The whole plot is divided into 5 different cross-sections representing each group-count. Fig. 3b represents 5 graphs between *different search algorithms* and *different type of street-networks* for different values of $|V^s|$. The black-boxed circle marks the lowest fractional error for that row. It is observed that for lower $|V^s|$, the D-Search algorithm gives lowest percentage error for most of the cities irrespective of their frac-Ds. However, this out-performance gets shifted towards R-Search for higher group-counts, with performance of similar intensity at $|V^s| = 5$ as that for D-Search at $|V^s| = 1$. It is, therefore, concluded that for complex E-GTSP scenarios it is better to select radially close points from each group with respect to the start-end points than to find points closest to their corresponding Dijkstra-route.

6. Conclusion

The E-GTSP, which is an extension of TSP, is proven by many researchers to better model real-life combinatorial optimization problems, where the task is to estimate open/close *Hamiltonian* cycle visiting each group exactly once in a given graph. A general approach is to reduce it to corresponding TSP before solving it to optimality. In this article, 5 different spatially driven search algorithms are presented and tested to transform E-GTSP to TSP. They have been tested out for 15 cities selected based upon the frac-D for 5 different $|V^s|$ instances each.

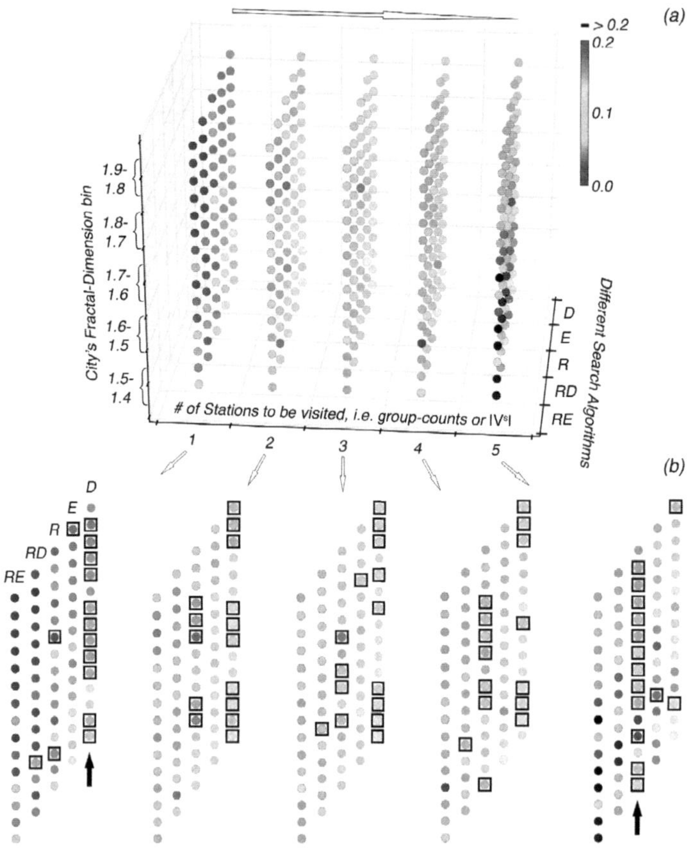

Figure 3: 3D-graph between *Different Search Algorithms*, *Number of Stations to be visited*, i.e. $|V^s|$, and *City's Fractal-Dimension bin*. Each bin represents 3 cities taken from Fig. 2. Color represents the average fractional error in route-length coming out from each algorithm with respect to the optimal one (estimated by Brute-Force), with black color for errors more than 20%. (a) It should be noted that with increase in $|V^s|$ for a given E-GTSP instance the error increases, irrespective of the choice of the algorithm (marked with big arrow). (b) Graph is sliced-down for each group-count, individually. For each city the algorithm with least error is marked with *black box*. It is clear that for small $|V^s|$ values the *D-Search* approach is better, and as one increases the number of stations the *R-Search* outperforms other, marked by solid-arrows.

It is observed that with increase in instance complexity all algorithms get erroneous, thereby giving longer routes which is independent of the choice of city street-pattern (Fig. 3a). For lower $|V^s|$ instances D-Search and for higher $|V^s|$ instances R-Search is better among all given algorithms (Fig. 3b). Overall, R-Search is the best approach, qualitative-wise, for tested cities by giving least route-length values. Its win could be attributed by its ability to consider stations belonging to regions outside the start-end proximity.

This study, thus, has brought forth a new search criterion of point/vertex selection from each group for an adequate E-GTSP to TSP transformation, tested on OSM dataset. This considers the radial spread of all points from each group with respect to the start/end point for optimal selection. It has widened up research possibilities in this pursuit; and researchers are encouraged to use tested dataset provided by Zia 2016 as benchmark for subsequent studies in order to escape long brute-force looping.

Future research will include testing R-Search criterion with other algorithms, like the state-of-the-art Lin-Kernighan-Helsgaun TSP solver, as discussed by researchers. Additional work might involve developing heuristic-R-Search or ANN-R-Search algorithm. Authors are optimistic that this attempt has added up few novel concepts to understand the old famous GTSP problem.

References

Bourke, P., 2003. Fractal Dimension Calculator - FDC. Online, online; accessed 15-November-2016. URL http://paulbourke.net/fractals/fracdim/

Clarke, G., Wright, J. W., August 1964. Scheduling of Vehicles from a Central Depot to a Number of Delivery Points. Operations Research 12(4), 568–581.

Croes, G. A., December 1958. A Method for Solving Traveling-Salesman Problems. Operations Research, 791–812.

Curtin, K. M., Voicu, G., Rice, M. T., Stefanidis, A., April 2014. A Comparative Analysis of Traveling Salesman Solutions from Geographic Information Systems. Transactions in GIS 18(2), 286–301.

Glover, F., August 1990. Tabu Search: A Tutorial. Interfaces, 74–94.

Gutin, G., Yeo, A., 2007. The Greedy Algorithm for the Symmetric TSP. Algorithmic Operations Research 2(1), 33–36.

Hamilton, W. R., December 1856. Memorandum respecting a new system of roots of unity. Philosophical Magazine 12, 446.

Helsgaun, K., 2000. An Effective Implementation of the Lin-Kernighan Traveling Salesman Heuristic. European Journal of Operational Research 126, 106–130.

Helsgaun, K., October 2009. General k-opt submoves for the LinKernighan TSP heuristic. Mathematical Programming Computation 1(2), 119–163.

Helsgaun, K., September 2015. Solving the equality generalized traveling salesman problem using the LinKernighanHelsgaun Algorithm. Operations Research Letters 7(3), 269–287.

Henry-Labor, A. L., 1969. The record balancing problem: A dynamic programming solution of a genera salesman problem. RAIRO - Operations Research 2, 43–49.

Kirkpatrick, S., Jr., C. D. G., Vecchi, M. P., May 1983. Optimization by Simulated Annealing. Science 220(4598), 671–680.

Laporte, G., Nobert, Y., 1983. Generalized Travelling Salesman Problem Through n Sets Of Nodes: An Integer Programming Approach. INFOR: Information Systems and Operational Research 21(1), 61–75.

Lenstra, J. K., Kan, A. H. G. R., November 1975. Some Simple Applications of the Travelling Salesman Problem. Operational Research Quarterly 26(4), 717–733.

Noon, C. E., Bean, J. C., August 1991. A Lagrangian Based Approach for the Asymmetric Generalized Traveling Salesman Problem. International Journal of Production Research 39(4), 623–632.

Psaraftis, H. N., Wen, M., Kontovas, C. A., January 2016. Dynamic vehicle routing problems: Three decades and counting. Networks 67(1), 3–31.

Rosenkrantz, D. J., Stearns, R. E., II, P. M. L., 2009. An analysis of several heuristics for the traveling salesman problem. Fundamental Problems in Computing 1, 45–69.

Saksena, J. P., November 1970. Mathematical model for scheduling clients through welfare agencies. Canadian Operational Research Society 8(3), 185.

Srivastava, S. S., Kumar, S., Garg, R. C., Sen, P., July 1969. Generalized travelling salesman problem through n sets of nodes. Canadian Operational Research Society 7(2), 97.

T., G. M., Roberto, T., 2015. The Christofides Approximation Algorithm. Algorithm Design and Applications 18(1), 513514.

Valenzuela, C. L., Jones, A. J., October 1997. Estimating the Held-Karp lower bound for the geometric TSP. European Journal of Operational Research 102(1), 157–175.

Zia, M., 2016. E-GTSP-to-TSP-tested-instances. Online, online; accessed 15-November-2016. URL https://github.com/Zia-/E-GTSP-to-TSP-tested-instances

Algorithm 1 ***E-Search*** & ***D-Search*** pseudo-code

Require: Terminal points S and F, $G = (V, E)$ representing urban street-network, & $V^g = \{V_1, V_2, ..., V_x\}$, i.e. points' groups set, where $V_i \subseteq V$ & $V_i \cap V_j = \varnothing \ \forall \ i, j = 1, 2, ..., x$ & $i \neq j$.

procedure E-GTSP $\Rightarrow$ TSP
  if Algorithm 1 $\widehat{=}$ ***E-Search*** **then** Draw SF **Euclidean** Line
  end if ▷ $\widehat{=}$ means corresponds to
  if Algorithm 1 $\widehat{=}$ ***D-Search*** **then** Calculate SF **Dijkstra** Route
  end if
  Initialize an empty container $C1$
  for $V_i \in V^g \ \forall \ i = 1, 2, .., x$ **do**
    Initialize an empty container $C2$.
    for $v_m \in V_i \ \forall \ m = 1, 2, .., |V_i|$ **do**
      $C2 \leftarrow v_m - SF$ shortest Euclidean distance ▷ $\leftarrow$ means append
    end for
    return $C2$
    $C1 \leftarrow v \ \widehat{=} \ v_m - SF \in C2$, with the smallest Euclidean distance
  end for
  return $C1$
end procedure
procedure TSP
  Calculate the minimum-cost *Hamiltonian* cycle induced by $C1$. *Note:* $C1 = V^s$.
end procedure

Development of an extension of GeoServer for handling 3D spatial data

Hyung-Gyu Ryoo[a,*], Soojin Kim[a], Joon-Seok Kim[a], Ki-Joune Li[a]

[a]*Department of Computer Science and Engineering, Pusan National University*

Abstract: Recently, several open source software tools such as CesiumJS and iTowns have been developed for dealing with 3-dimensional spatial data. These tools mainly focus on visualization of 3D spatial data based on WebGL. An open-sourced server capable of storing, sharing and querying 3D spatial data has not yet been developed. GeoServer, one of the representative open source spatial data servers, provides many powerful features. In particular, it supports connecting to and publishing spatial data from a variety of data sources. GeoServer also supports Web Feature Service (WFS), which is a standard protocol established by the Open Geospatial Consortium to request geospatial feature data. However, GeoServer provides functions only for two-dimensional geometry, so it provides few functions for handling 3D spatial data. Because JTS Topology Suite, which is an important component of GeoServer, does not support 3D spatial operations, it also does not support solid geometries. In this paper, we discuss extension modules of GeoServer that we have implemented to handle 3D spatial data. First, instead of JTS, our modules adopted a geometry model based on the ISO 19107 standard and support 3D spatial operations from the Simple Feature CGAL library. Based on this geometry model, we have implemented new internal data structures that represent spatial information from the Feature interface in GeoServer. Second, we also extended the DataStore module to handle and store 3D spatial information for several data sources such as GeoJSON, GML and PostGIS. Finally, we extended the WFS module to share 3D spatial data via GeoServer.

*Corresponding author
Email address: `hgryoo@pnu.edu` (Hyung-Gyu Ryoo)

1. Introduction

Recently, several open source software tools dealing with 3D spatial data such as CesiumJS (AGI 2017) and iTown (Oslandia 2016) have been actively developed. Most of them focus mainly on visualizing 3D spatial data in a client environment. However, a universal spatial data server for providing 3D spatial data to such a visualization tool has not yet been developed, and therefore it is hard for the client to service 3D spatial information. In order to provide geospatial web services to the users, it is necessary to develop a spatial data server able to share and query 3D spatial data.

GeoServer (GeoServer 2001) is a representative open source spatial data servers that provide a variety of spatial data sources that conform to open standards. It supports file formats for handling spatial data such as a GeoJSON (Butler et al. 2008) and a shapefile (ESRI 1998) and geospatial database management systems (DBMS) such as PostGIS (PostGIS 2001) and Oracle Spatial and Graph (Oracle 2016). GeoServer allows users to add new data stores as a form of plugin. Based on these various spatial data sources, GeoServer is an important component in geospatial web services that provides functions to share and query spatial data by implementing open geospatial web services established by Open Geospatial Consortium (OGC) such as Web Map Service (Service 2006), Web Feature Service (WFS/FES 2005) and Web Processing Service (Service 2015).

Although it is powerful as a spatial data server for general purposes, GeoServer has limitations in handling 3D spatial data. For instance, it is impossible to store spatial information with solids, which are bounded by closed surfaces, into data stores as well as to load that from data sources. It is possible for GeoServer to handle and store points, curves, and surfaces with three-dimensional coordinates. However, GeoServer barely supports 3D spatial queries because it only performs spatial query processing that ignores z coordinates.

The limitations originate from data structures representing spatial information in GeoServer. JTS Topology Suite (Java Topology Suite), a fundamental library written in Java, (Tech 2000) is a core component of GeoServer for geometries. On the top of JTS, GeoServer provides a number of spatial operations and utility functions for various 2D geometries. In the most implementations, it is assumed that geometry attributes of a feature are represented as a geometry of JTS library. In other words, GeoServer is deeply coupled with JTS. Therefore, it is necessary to replace JTS for GeoServer to provide fully functional features in 3D.

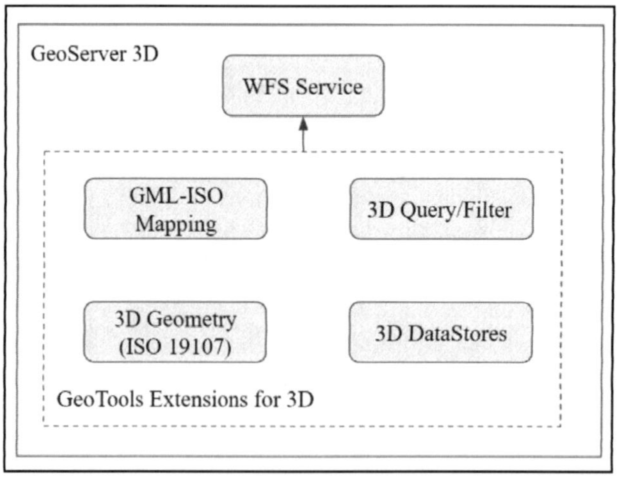

Figure 1: Overview of our extension of GeoServer to support 3D spatial data

In this paper, we introduce our extension of GeoServer to support 3D spatial data as shown in Figure 1 (STEMLab 2015, STEMLab 2016). Our contribution to GeoServer and open source communities is summarized as follows:

- We have taken the initiative to implement a fully functional open-sourced data server for 3D spatial data.
- We carefully examined GeoServer from bottom to top to determine how to add a 3D extension.
- We identified requirements and issues for a 3D spatial data server based on GeoServer.
- We built a feasible full stack composed of open source software.

This paper is organized as follows. We survey related work and present requirements for a 3D spatial data server in Section 2. In Section 3, we introduce implementations of data structures by using geometric models based on ISO 19107 spatial schema (ISO/TC211 2003) to store 3D geometries in GeoServer. Section 4 describes the data store that can store and query 3D geometries based on the data structure implemented in Section 3. In Section 5, we describe support for the Web Feature Service protocol to share 3D spatial information. In Section 6, we conclude the paper and discuss current limitations and future work.

2. Related Work

There are several DBMSs that can store 3D spatial data and perform query processing on spatial information. Oracle Spatial and Graph, a representative commercial database, supports most spatial operations and 3D spatial data types, including a solid geometry. PostGIS, a widely used open source spatial database, partially supports 3D spatial data types and operations. Although spatial DBMSs are fundamental to manipulate spatial data, common interfaces and tools have been demanded to satisfy various geospatial applications. For instance, users may need to overlay and manipulate multiple layers from various data sources without concerns about different implementations on each data store.

GeoTools (GeoTools 1998) is an open source library that meets these requirements. It provides excellent abstraction for multiple data sources. GeoServer is an implementation of a geospatial data server based on GeoTools. It allows users to simply load/store layers from/to various databases and file-based data sources, hiding details of how to handle them. A readable/writable source is called a data store. Thanks to many contributors, GeoServer supports a variety of data stores. Also, it lets users to add data stores via the Web. The reason is that GeoServer has a common interface, called Feature, between data stores and a module that provides the Web Feature Service. Thus, it acts as a bridge between heterogeneous data stores and the user interface, to exchange the spatial information.

A Feature instance can have spatial and non-spatial properties. For geometric properties, GeoServer employs JTS, a widely used open source library written in Java. Since it is robust, there exist ports of JTS in different languages such as Geometry Engine, Open Source (GEOS 2003). JTS implements the OpenGIS Simple Features Specification, which is designed to handle 2D geometry types such as points, curves and surfaces and operations with them. Unfortunately, most GeoServer modules including Feature are deeply coupled with the JTS. In order to provide 3D spatial operations in GeoServer, it is necessary to replace the geometry properties of Feature and relevant modules. If GeoServer itself has 3D spatial operation capabilities, clients can store and query 3D spatial data to any data stores.

There exists a 3D geometry model in GeoTools as an unsupported module, gt-geometry, and it is based on ISO 19107 spatial schema. Due to the complexity of 3D spatial operations, it has had rare application. Therefore, there remains a problem of how to provide 3D spatial operations when using the ISO 19017-based geometric model as the value of the geometry property of Feature.

For implementing 3D spatial operations, there may be two options: 1) to develop or port every function from scratch and 2) to reuse or utilize existing open source libraries. Although it could be optimized for GeoServer, the first option requires a lot of effort to make the functions robust and efficient and the operations should be maintained separately. In this paper, we focus on the second method to take advantage of open source libraries that provide enough functionality.

Computational Geometry ALgorithms (CGAL) is an open source library that provides a variety of robust computational geometry algorithms in 3D spaces as well as 2D (Fabri and Pion 2009). Since it is designed for general purposes, the geometry model and operations in CGAL are different from those in GIS domains. There is an existing open source project to that fills the gap called Simple Feature CGAL (Oslandia1 2016). This library is implemented using CGAL and Boost. Simple Feature CGAL (SFCGAL) implements a geometric model and application programming interfaces (APIs) defined in the ISO 19107 spatial schema and the OGC Simple Features Access 1.2 standard (Features 2011). SFCGAL is robust and yields highly precise results. There is a trade-off between precision and performance when using CGAL.

From the survey of existing open source projects, we identify requirements of a 3D spatial data server as follows:

- Support data structures to store 3D geometries
- Support connections to data stores able to store 3D spatial information
- Support functions to process queries on 3D objects
- Provide standard web services for sharing 3D spatial information

3. Implementation of Data Structure for 3D Geometry Model

GeoServer abstracts all spatial information into a class called Feature and the geometry attributes of spatial information are represented as GeometryAttribute. The GeometryAttribute interface has the actual geometry object implemented in Java. In most implementations, it is assumed that geometry attributes of a feature are represented as a geometry of the JTS library. The problem is that JTS geometry objects are represented not as a Java interface but as a Java class. This prohibits other implementation, making it difficult to extend the functionality. Therefore, it is necessary to replace the JTS and relevant modules with other libraries.

First, an interface-based structure is required for extensibility. Currently, an unsupported module in GeoServer defines and partially implements an ISO 19107 spatial schema-based interface. It has the Solid interface but no class implementing it. Also, the module does not support 3D spatial operations such as Contains and Intersects for two geometries with Z coordinates. Few functions such as finding the distance between geometries are implemented. Therefore, we added Shell class and Solid class to be compatible with ISO 19107.

In order to support 3D spatial operations, we have connected the SFCGAL library to the existing interface of the ISO 19107 spatial schema. Figure 2 shows its structure. Since SFCGAL is a library written in C ++, a bridge between native C++ and Java is required to invoke

SFCGAL functions. Among the libraries that support this, we adopted JavaCPP, which is an open source tool and relatively easy to interface. JavaCPP can generate wrapper classes corresponding to SFCGAL. The structure of SFCGAL geometry is the same as Simple Feature Geometry.

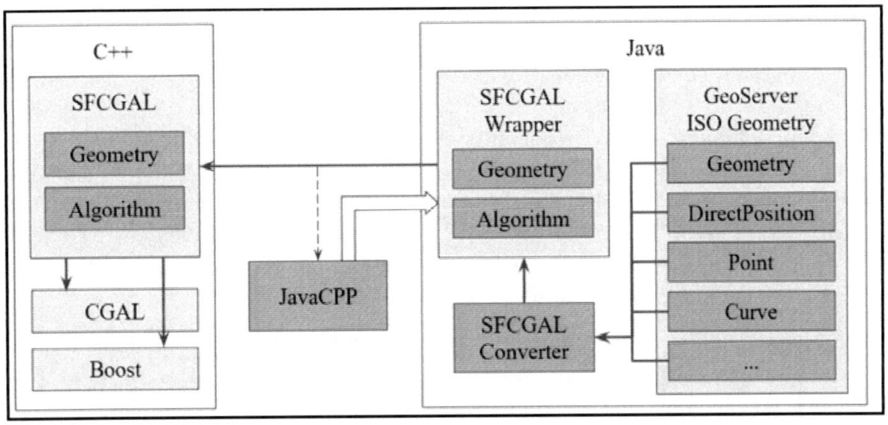

Figure 2: Connecting SFCGAL to support 3D operations

In order for communication between SFCGAL and GeoServer, conversion process is necessary because of a gap between the SFCGAL Wrapper classes and the GeoServer ISO Geometry interface (see Figure 2). In our extension, we implemented a converter that converts from ISO geometry to SFCGAL and vice versa. Then, SFCGAL Wrapper objects invoke the corresponding native methods of SFCGAL when methods of ISO geometry are called by GeoServer. Once processing in SFCGAL is completed, the results are returned as wrapper objects. The mapping relation between Simple Feature Access, SFCGAL and ISO geometry is defined as Table 1.

<table>
<tr><th>Simple Feature Access</th><th>SFCGAL</th><th>ISO 19107</th></tr>
<tr><td>Coordinate</td><td>Coordinate</td><td>DirectPosition</td></tr>
<tr><td>Point</td><td>Point</td><td>Point</td></tr>
<tr><td>LineString</td><td rowspan="3">LineString</td><td rowspan="2">Curve</td></tr>
<tr><td>Line</td></tr>
<tr><td>LinearRing</td><td>Ring</td></tr>
<tr><td>Polygon</td><td>Polygon</td><td rowspan="2">Surface</td></tr>
<tr><td>Triangle</td><td>Triangle</td></tr>
<tr><td rowspan="2">PolyhedralSurface</td><td>PolyhedralSurface</td><td>PolyhedralSurface</td></tr>
<tr><td>TriangulatedSurface</td><td>TriangulatedSurface</td></tr>
<tr><td>TIN</td><td>TIN</td><td>TIN</td></tr>
<tr><td>-</td><td>Solid</td><td>Solid</td></tr>
<tr><td>MultiPoint</td><td>MultiPoint</td><td>MultiPoint</td></tr>
<tr><td>MultiLineString</td><td>MultiLineString</td><td>MultiCurve</td></tr>
<tr><td>MultiPolygon</td><td>MultiPolygon</td><td>MultiSurface</td></tr>
<tr><td>-</td><td>MultiSolid</td><td>MultiSolid</td></tr>
<tr><td>GeometryCollection</td><td>GeometryCollection</td><td>MultiPrimitive</td></tr>
</table>

Table 1: Mapping relation between Simple Feature Access, SFCGAL and ISO geometry

4. A Data Store for 3D Spatial Information

In order to support a general purpose 3D spatial information data server, it is necessary to provide various data stores capable of storing 3D spatial information. In GeoServer, data stores can be divided into two types: 1) a file data store and 2) a database data store. There are file data stores supporting file formats such as Shapefile, GeoJSON, and database data stores manipulating databases such as Oracle, PostGIS, and MySQL.

This section describes two data stores for GeoJSON, a spatial file format with a relatively simple structure, and PostGIS, a widely used open source spatial database. For PostGIS, if you use the SFCGAL extension, you can use geometry with Z coordinates and 3D operations. In GeoServer, each concrete store can be extended to the plug-in by implementing the necessary interfaces to the store. Likewise, new stores can be extended based on the methods described in this section.

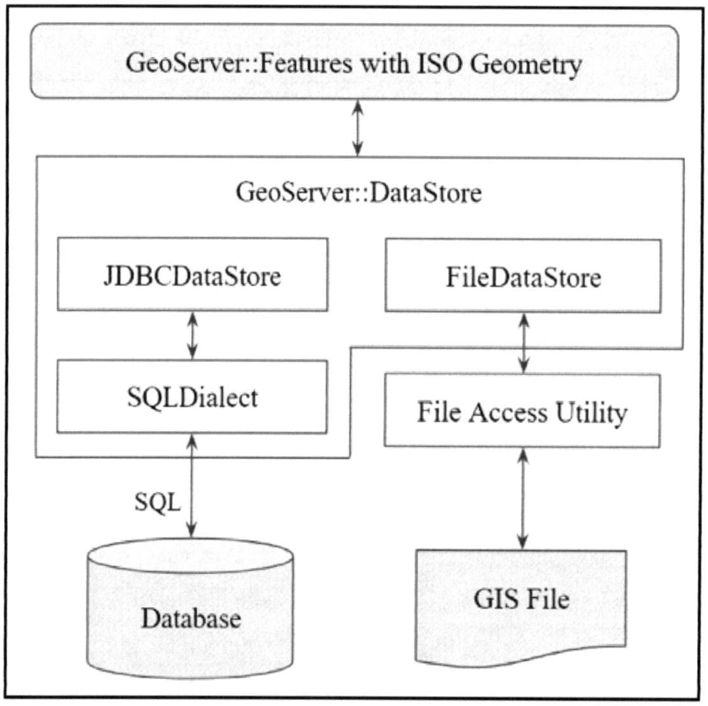

Figure 3: DataStore architecture in GeoServer

GeoServer provides a function that supports data store by implementing an interface called DataStore as shown in Figure 3. In GeoServer, DataStore is an abstraction of persistent object stores containing instances of Feature. Any implementations of DataStore are responsible for importing/exporting features from/to their physical stores. File data stores should be an implementation of the FileDataStore interface to support their file formats. Although they have different capabilities, DBMSs with a Java Database Connectivity (JDBC) driver have common functionality. For the common parts, GeoServer provides an abstract class called JDBCDataStore. Each database data store needs to extend SQLDialect for its own spatial types and functions.

4.1. Implementation of DataStore

GeoJSON is an open standard format that supports simple features, based on JavaScript Object Notation (JSON), which is mainly used for data exchange in the web environment. GeoJSON is a lightweight data format, compared to rich formats based on XML such as GML. It supports simple geometry types: Point, LineString, Polygon, MultiPoint, MultiLineString, MultiPolygon, and GeometryCollection. In order to take advantages of GeoJSON, we define an extended GeoJSON for solids by adding Solid. Structure of Solid's coordinates is similar to MultiPolygon and Figure 4 is an example of cube in extended GeoJSON.

```
{ "type": "FeatureCollection",
  "features": [
  { "type": "Feature",
  "geometry":
    { "type": "Solid",
    "coordinates": [[
      [[0.0,0.0,0.0],[0.0,0.0,1.0],[0.0,1.0,1.0],[0.0,1.0,0.0],[0.0,0.0,0.0]],
      [[0.0,0.0,0.0],[0.0,1.0,0.0],[1.0,1.0,0.0],[1.0,0.0,0.0],[0.0,0.0,0.0]],
      [[0.0,0.0,0.0],[1.0,0.0,0.0],[1.0,0.0,1.0],[0.0,0.0,1.0],[0.0,0.0,0.0]],
      [[1.0,1.0,0.0],[1.0,1.0,1.0],[1.0,0.0,1.0],[1.0,0.0,0.0],[1.0,1.0,0.0]],
      [[0.0,1.0,0.0],[0.0,1.0,1.0],[1.0,1.0,1.0],[1.0,1.0,0.0],[0.0,1.0,0.0]],
      [[0.0,0.0,1.0],[1.0,0.0,1.0],[1.0,1.0,1.0],[0.0,1.0,1.0],[0.0,0.0,1.0]]
    ]]},
    "properties": {
     "name": "Room A",
     "floor": 1
     }}]}}
```

Figure 4: Example of Solid in the extended GeoJSON

We implemented a new FileDataStore to handle the extended GeoJSON. While they parse/interpret files in specific formats, some file data stores without schema definitions including GeoJSON define types of features and instantiate them. For consistency, geometry attributes of features in the extended GeoJSON are interpreted as ISO geometry types in GeoServer according to Table 1. This enables users to import/export features with a solid geometry from/to the extended GeoJSON.

In a database data store, a feature type is created from table schema because records in a conventional DBMS such as PostGIS are shaped by schema. Because PostGIS provides geometric types based on Simple Feature Access, geometry attributes of features can be represented as ISO geometry according to Table 1.

PostGIS supports 3D geometry types and operations if and only if it has the SFCGAL extension. This extension includes additional types such as PolyhedralSurface and Solid. It also provides various operations that take into account Z coordinates. To maximize utilization of PostGIS capabilities, the data store is implemented using the SFCGAL extension. As a result, this reduces gaps between two data models so we could avoid a strained implementation by converting solids into alternative types such as MultiSurface. On the SFCGAL installation, the data store enables users to import/export features with 3D geometries from/to PostGIS.

4.2. 3D Spatial Query Processing

This section describes how to process spatial queries for each data store. GeoServer defines the query conditions based on the OGC Filter Encoding Specification (WFS-FE 2005). Therefore, queries to the DataStore are implemented to pass the filter condition and process it. Figure 5 shows the overall flow to perform query processing with 3D spatial filters.

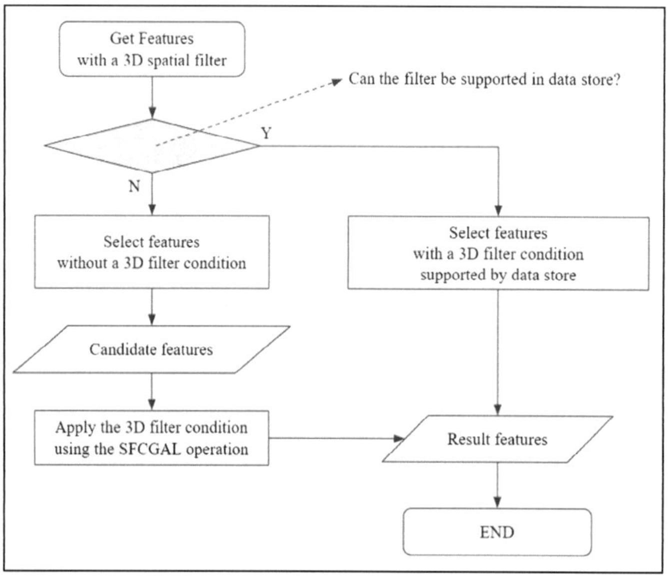

Figure 5: Flow diagram of handling a query with 3D spatial filter in data store

Basically, 3D spatial queries including 3D spatial filters can be handled without assistance of a data store in our extension of GeoServer. A spatial filter such as Contains and Intersects, however, can considerably reduce the number of candidates if a data store supports the filter. As file stores have no spatial operation capability, any 3D spatial filter will be processed using the SFCGAL. On the other hand, database data stores can apply some spatial filters, depending on their capabilities.

5. Web Feature Service Extension for 3D Queries

Web Feature Service (WFS), one of OGC's standard web protocols, is an interface for providing spatial data in vector format. This allows performing data manipulation operations including retrieving, creating, deleting, and updating features across the web. In this paper, we implemented the functionality of Web Feature Service 1.1 to access, store, query, and share 3D spatial information through the web environment. WFS 1.1 supports the GML 3.1.1 used in HTTP requests from and responses to the client. Since its geometry model is an implementation of ISO 19107 in XML, the GML 3.1.1 contains solids.

However, GeoServer does not give a correct response when the WFS client makes a request to the store containing features with solids. The reason is that a GML geometry is interpreted as a geometry of JTS while parsing GML instances. Therefore, in order to work properly, we implemented a module mapping each GML geometry not to a geometry of JTS but to a geometry defined in ISO 19107.

WFS requests are handled according to the flow as shown in Figure 6. WFS requests sent by a WFS client to GeoServer include an XML-formatted Filter and GML. The WFS function of GeoServer converts GML, the geometry information described in the request statement, to the geometry type used by GeoServer. For example, if you have a 3D region as a query condition, the 3D region in GML is represented by gml:Solid, which is converted to the solid type of the ISO geometry. Also, the features that result from the query performed in the data store will contain a 3D geometry type. This 3D geometry type is expressed in ISO geometry. Therefore, we converted the ISO geometry of each feature to GML to create a WFS response.

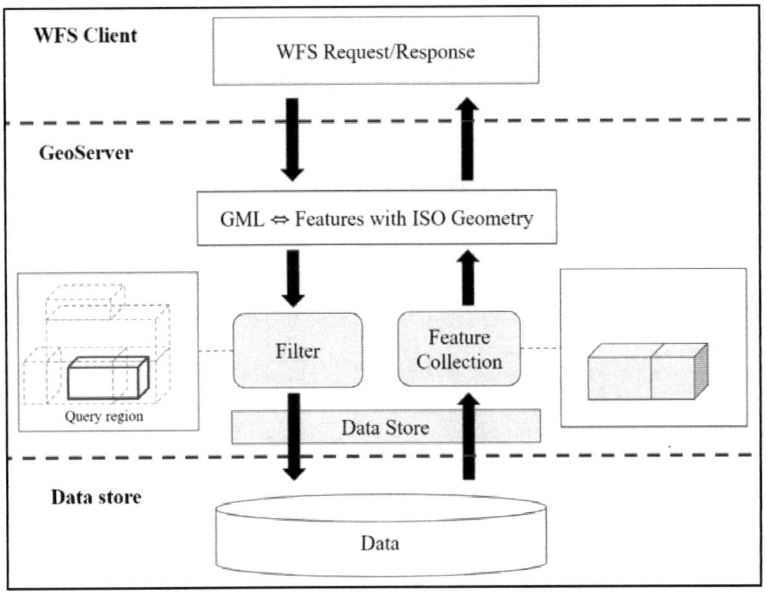

Figure 6: Flow of converting between GML and Feature with ISO Geometry

As a result, Figure 7 visualizes the result of a GetFeature query by implementing a simple WFS client using CesiumJS. The example data of 3D objects is composed of rooms and corridors in a shopping mall in Seoul. The number of rooms is 4893 and each room is made of solid geometry. In Figure 8, the client sends a query using the WFS 1.1 GetFeature to find the overlapping room by specifying a query area with a 3D space represented by the red portion. As a result, the request is executed in the 3D store where the 3D objects are stored, and each room corresponding to the query result is returned as the Figure 9, and the corresponding geometry is displayed in yellow in the WFS client.

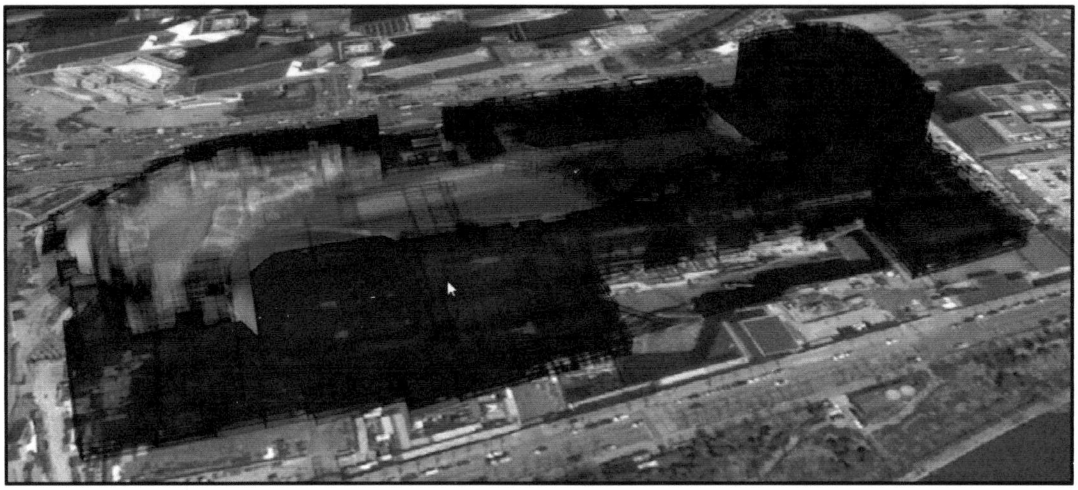

Figure 7: A result of GetFeature operation selecting features with Solid geometry

```
<?xml version="1.0" ?>
<wfs:GetFeature
        service="WFS"
        version="1.1.0"
        outputFormat="text/xml; subtype=gml/3.1.1"
        xmlns:stem="http://stemlab.pnu.edu"
        xmlns:wfs="http://www.opengis.net/wfs"
        xmlns:ogc="http://www.opengis.net/ogc"
        xmlns:xsi="http://www.w3.org/2001/XMLSchema-instance"
        xsi:schemaLocation="http://www.opengis.net/wfs ../wfs/1.1.0/WFS.xsd">
        <wfs:Query typeName="stem:solid3d_sfcgal">
          <ogc:Filter>
            <ogc:Intersects>
              <ogc:PropertyName>stem:geom</ogc:PropertyName>
              <gml:Envelope srsName="EPSG:4329">
                <gml:lowerCorner>127.10156397 37.46185285 -100</gml:lowerCorner>
                <gml:upperCorner>127.10226404 37.46227915 100</gml:upperCorner>
              </gml:Envelope>
            </ogc:Intersects>
          </ogc:Filter>
        </wfs:Query>
</wfs:GetFeature>
```

Figure 8: Example of the GetFeature request with Intersects query

```
<?xml version="1.0" encoding="UTF-8"?>
<wfs:FeatureCollection
        xmlns:xs="http://www.w3.org/2001/XMLSchema"
  xmlns:wfs="http://www.opengis.net/wfs"
  xmlns:gml="http://www.opengis.net/gml"
  xmlns:ogc="http://www.opengis.net/ogc"
  xmlns:ows="http://www.opengis.net/ows"
  xmlns:xlink="http://www.w3.org/1999/xlink"
  xmlns:stem="http://stemlab.pnu.edu"
  xmlns:xsi="http://www.w3.org/2001/XMLSchema-instance"
  numberOfFeatures="309" timeStamp="2017-05-21T13:25:58.214Z"
  xsi:schemaLocation="http://www.opengis.net/wfs
  http://127.0.0.1:8883/geoserver/schemas/wfs/1.1.0/wfs.xsd http://stemlab.pnu.edu
  http://127.0.0.1:8883/geoserver/wfs
  service=WFS&amp;version=1.1.0&amp;
  request=DescribeFeatureType&amp;typeName=stem%3Asolid3d_sfcgal"
  ...
  <gml:featureMembers>
  <stem:solid3d_sfcgal gml:id="solid3d_sfcgal.fid--4a5ecf22_15c2b2204b3_-7ff7">
  <gml:boundedBy>
        <gml:Envelope srsName="urn:x-ogc:def:crs:EPSG:4329" srsDimension="3">
                <gml:lowerCorner>127.10227883229379 37.51336579829316 0.0</gml:lowerCorner>
                <gml:upperCorner>127.10479641589909 37.514514506176944 3.0</gml:upperCorner>
  </gml:Envelope></gml:boundedBy>
  <stem:fid>C835</stem:fid>
  <stem:geom>
        <gml:Solid srsName="urn:x-ogc:def:crs:EPSG:4329" srsDimension="3">
                <gml:exterior><gml:CompositeSurface><gml:surfaceMember>
                <gml:Polygon><gml:exterior><gml:LinearRing>
                <gml:posList>
        127.10233736459824 37.51337631034256 3.0 127.10231000709786 37.51342579228809 3.0
        127.10227883229379 37.51341528023869 3.0 127.10230618979418 37.51336579829316 3.0
        127.10233736459824 37.51337631034256 3.0</gml:posList>
                </gml:LinearRing></gml:exterior></gml:Polygon>
                </gml:surfaceMember>
                        ...
                </gml:CompositeSurface>
                </gml:exterior>
        </gml:Solid>
        </stem:geom>
        </stem:solid3d_sfcgal>
        ...
</gml:featureMembers>
</wfs:FeatureCollection>
```

Figure 9: Example of the GetFeature response containing features with Solid geometry

6. Conclusion

GeoServer has been widely used as an open source spatial data server. However, GeoServer does not support 3D spatial data handling because it has been implemented based on Simple Feature Access Specification. Therefore, this study used the ISO 19107 geometric model which can support 3D spatial geometry and computation. Based on this, we extended GeoServer to support the data stores that can store 3D spatial information. We also extended GeoServer's WFS interface to share and query the spatial information stored in the 3D data store across the web. The version of GeoServer that is developed from this study can be used in fields requiring 3D spatial information and queries. For example, if there is a complicated pipe in a building, it would be possible to query the rooms where the pipe passes.

However, there are still items to improve. Currently, as a 3D spatial server, GeoServer does not provide all kinds of 3D data types and operations for general purposes. For example, the SFCGAL library does not support curve geometries like arcs or geodesics, so we cannot support all geometries in ISO 19107. Thus, it is an interesting topic to implement a function to provide curved geometry. It will be practical if our next version can support WFS 2.0, the latest version of Web Feature Service, which provides rich functionality by supporting temporal queries and join queries through GetFeature requests. It also supports complex filter queries that can be stored in WFS server. Furthermore, 3D Portrayal IE (Portrayal 2015), a web-based interface for supporting 3D scene rendering at the server side, is being actively discussed as a standard candidate for OGC. It is necessary to provide a server capable of handling 3D spatial information to support such a standard.

References

AGI, 2017. Cesium webgl virtual globe and map engine. Online, online; accessed May 23, 2017. URL `http://cesiumjs.org`

Butler, H., Daly, M., Doyle, A., Gillies, S., Schaub, T., Schmidt, C., 2008. The geojson format specification. Rapport technique 67.

ESRI, 1998. Shapefile technical description. An ESRI White Paper 7855.

Fabri, A., Pion, S., 2009. Cgal: The computational geometry algorithms library. Proceedings of the 17th ACM SIGSPATIAL international conference on advances in geographic information systems, 538–539.

Features, O. S., 2011. Opengis implementation specification for geographic information - simple feature access - part 1: Common architecture. Online, online; accessed May 23, 2017. URL `http://www.opengeospatial.org/standards/sfa`

GEOS, 2003. Geos geometry engine, open source. Online, online; accessed May 23, 2017. URL `https://trac.osgeo.org/geos`

GeoServer, 2001. Geoserver. Online, online; accessed May 23, 2017. URL `http://geoserver.org`

GeoTools, 1998. Geotools the open source java gis toolkit. Online, online; accessed May 23, 2017. URL `http://geotools.org`

ISO/TC211, 2003. Geographic information - spatial schema, iso. An ESRI White Paper 19107:2003.

Oracle, 2016. Oracle spatial and graph 12c release 2. Online, online; accessed May 23, 2017. URL `http://www.oracle.com/technetwork/database-options/spatialandgraph/overview/spatialandgraph-1707409.html`

Oslandia, 2016. Oslandia itowns. Online, online; accessed May 23, 2017. URL `http://www.itowns-project.org`

Oslandia1, 2016. Simple feature cgal (sfcgal). Online, online; accessed May 23, 2017. URL `http://www.sfcgal.org/`

Portrayal, O. D., 2015. 3d portrayal service candidate standard. Online, online; accessed May 23, 2017. URL `http://www.opengeospatial.org/projects/initiatives/3dpie`

PostGIS, 2001. Postgis. Online, online; accessed May 23, 2017. URL `http://www.postgis.net`

Service, O. W. M., 2006. Opengis web map service (wms) implementation specification, 06-042. Online, online; accessed May 23, 2017. URL `http://opengeospatial.org/standards/wms`

Service, O. W. P., 2015. Wps 2.0 interface standard, 04-065. Online, online; accessed May 23, 2017. URL `http://opengeospatial.org/standards/wps`

STEMLab, 2015. Geotools 3d extension. Online, online; accessed May 23, 2017. URL `https://github.com/STEMLab/geotools-3d-extension/`

STEMLab, 2016. Geotools 3d extension. Online, online; accessed May 23, 2017. URL `https://github.com/STEMLab/geoserver-3d-extension/`

Tech, L., 2000. Jts topology suite (java topology suite). Online, online; accessed May 23, 2017. URL `https://www.locationtech.org/projects/technology.jts`

WFS-FE, O., 2005. Opengis filter encoding implementation specification, 04-095. Online, online; accessed May 23, 2017. URL `http://www.opengeospatial.org/standards/filter`

WFS/FES, O., 2005. Opengis web feature service (wfs) implementation specification, 04-094. Online, online; accessed May 23, 2017. URL `http://opengeospatial.org/standards/wfs`

Exploring the Relationship Between Climate and Forest Conditions in Forest Classee De La Mondah (GABON) using Remote Sensing Data

Mariano MBOUMBA[a,*], Ichio ASANUMA[a], Keitarou HARA[a]

[a]*Departement of Informatics, Tokyo University of Information Sciences 4-1 Onaridai, Wakaba-ku, Chiba, 265-8501 Japan*

Abstract: This study aims to examine the relationship between temperature and precipitation as climatic factors and the vegetation characteristics in Forêt Classée de la Mondah (Gabon). A pixel based spatial analysis procedure with QGIS was used to extract vegetation characteristics from MODIS on Terra, NDVI band, data from February 2000 to December 2015, while temperature and precipitation were monitored from January 1980 to December 2015 to highlight climate fluctuations in the study area. Two regression analyses were conducted : one between temperature and NDVI and the other one between precipitation and NDVI. Results showed that temperature and precipitation did not vary significantly and the overall linear correlation between climate factors and NDVI was not significant. However, a close to medium correlation was found between temperature and NDVI during the long dry season. The authors are conducting a similar study in a broader scale (at country level), based on polygon as a spatial analysis procedure to extract MODIS data, in order to obtain further information about the relationship between climate and vegetation in this region and highlight the fact that vegetation may respond differently to climate effects depending on different micro climate.

*Corresponding author
Email address: `g16005mm@edu.tuis.ac.jp` (Mariano MBOUMBA)

1. Introduction

Climate change and its impacts on vegetation are a subject of special attention that is characterized by numerous publications on this topic. It results, among others, by a significant increase in global mean temperature, especially in higher latitudes, and by a change of rainfall regime causing heavy rainfall in some areas and dry and hot periods in others.

A climatic variability implies modifications of its indicators throughout time. Previous researches showed that major modifications of climate indicators occured in most of tropical regions. Climatic factors such as precipitation and temperature are directs and proximal factors influencing vegetation distribution. That is, either they have a known direct influence or they are considered to have a direct physiological influence on vegetation species, the most causal variables exhibit physiological responses (Hao et al. 2012; Watling et al. 2013). Previous studies have shown the influence of climate on forest ecosystem processes, among other factors, by the response of vegetation to climatic variables through the application of remote sensing, which proved to be a good tool for the description of vegetation dynamic (Okin et al. 2001).

This study aims to describe an approach on how geospatial tools can be applied in the sciences while examining the relationship between temperature and precipitation, used as climatic variables, and the vegetation characteristics in Forêt Classée de la Mondah (FCM, in Gabon). For this purpose, the free and open source software (FOSS) QGIS is used for processing and extracting satellite time series data in order to conduct regression analysis and determine the correlation between the variables mentioned above.

2. Material and Method

2.1. Study area

FCM is a forest dedicated for scientific researches of the National College of Water and Forest of Gabon, under the administrative authority of National Agency of National Parks of Gabon. Gabon, located in the west central Africa, is a country covered 85% of tropical rainforest. There are 4 seasons : two rainy seasons, long (February ∼ May, warm weather with lots of precipitations and a cloudy sky) and short (September ∼ November, warm weather with lots of precipitations and a cloudy sky); and two dry seasons, long (June ∼ August, cool weather with a negligible rate of precipitation and a clear sky) and short (December ∼ January, few precipitations, pretty warm weather and a cloudy sky). The total annual precipitations are from 2000 mm to 3800 mm while the monthly mean temperature is around 22 and 29. Its climate is classified to the Equatorial of transition.

FCM has an area of 102 km^2 and spans about 10 kilometers from 9°18' to 9°24' East longitude, and approximately 11 km from 0°37' to 0°29' North latitude. Figure 1 below shows the location in the country and administrative boundaries of the study area.

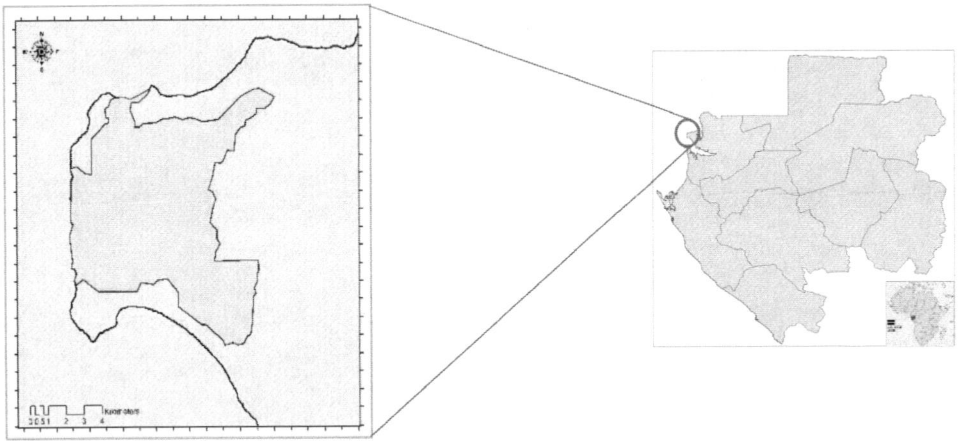

Figure 1: Location of Forêt Classée de la Mondah (FCM). The map on the right side locate FCM in the country while the map on left side shows the limits of FCM.

2.2. Data source and preparation

MODIS/Terra images were used in this study. A total of 191 MODIS/Terra monthly composite NDVI images (MOD13A3, NDVI) were downloaded from Land Processes Distributed Active Archive Center (LP DAAC). Mean temperatures and precipitations of the study area were acquired from National Oceanic and Atmospheric Administration (NOAA) and cover a period of 35 years, from 1980 to 2015. In order to remove seasonal changes and get a more accurate analysis in NDVI, temperature and precipitations time series, anomaly was calculated for every considered period in the datasets. It is worthwhile to note that anomaly is the departure from reference value (long-term average of a considered period). In case of temperature, for instance, a positive anomaly indicates that the observed temperature was warmer than the reference value, while a negative anomaly indicates that the observed temperature was cooler than the reference value (NOAA 2017; Anonymous 2017).

2.3. Selecting appropriate locations for extracting pixels

In order to emphasize the photosynthetic activity and exclude the potential effects of anthropogenic activities during the regression analysis process, a pixel based spatial analysis procedure with QGIS was used so that any NDVI value less than 0.156 over the whole time series was excluded from the analysis. These excluded NDVI values were representative of bare soil (Mollalo et al. 2014). Raster product tool allowed to create a binary raster (mask) of desired pixels in addition to delineate the targeted area in every single raster by multiplying the so created mask by the original time series. In this last case raster product tool is run as batch processing. In order to avoid erroneous results, the same spatial reference (WGS 84, UTM zone 32N, datum D WGS 1984) was set to the whole time series using Warp reproject tool and to the polygon of the study area as well. Then, satellite data were clipped and a mask created on the fly using raster calculator (from processing toolbox) run as batch processing (figure 2).

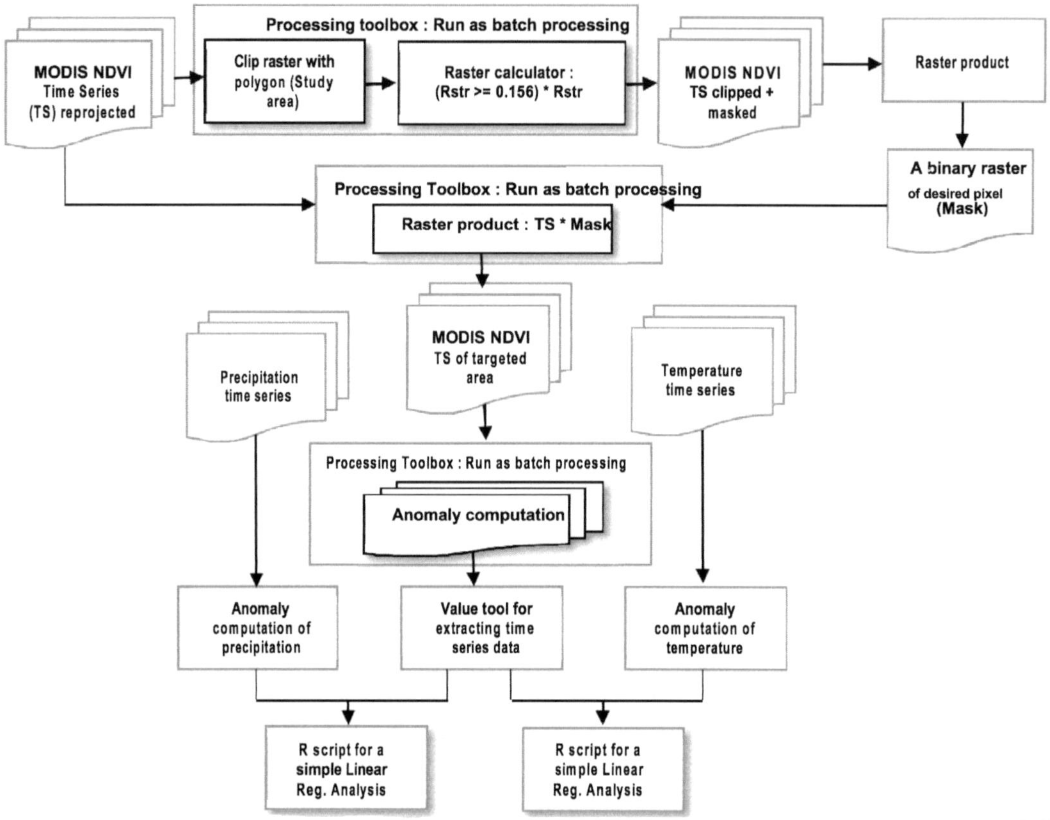

Figure 2: Approach for locating the targeted area and extracting time series data for the regression analysis using R script.

2.4. Approach for regression analysis between climate variables and NDVI

In order to account for the overall fluctuations of climate in the study area, an analysis of the trend of monthly mean temperature and precipitation anomalies was performed from 1980 to 2015. Then, temperature and precipitation time series were matched with NDVI time series, in order to conduct the regression analysis from February 2000 to December 2015. After anomaly was calculated, pixels were extracted from the targeted area (Figure 2) of the whole NDVI time series by using the Value tool plugin. Then, two linear regression analyses (One between temperature and NDVI and the other one between precipitation and NDVI) were conducted using R script so as to model the statistical relationship between climate variables and NDVI according to equation 1:

$$y = a + bx \tag{1}$$

Where a and b are the slope and intercept; x is independent (or explanatory) variables representing temperature or precipitation while y is dependent (or response) variable, representing NDVI.

To perform the analysis through R script, data were imported and attached using the commands read.delim and attach respectively. Then, after plotted scatter plot included the regression line (plot and abline commands), the linear regression was fitted to data using the lm command and the summary of the model was returned, displaying relevant information for the analysis such as the slope and intercept, the residual and residual standard error, R squared and adjusted R squared... The confidence interval for statistic test was set to 0.95 using the confint command. Finally, some regression diagnostic plots (residual plot, QQ plot, ...) were produced in order to check the validity of the model.

3. Results

3.1. Fluctuations of climate variables over 35 years

The general trend of temperature and precipitation anomalies did not vary significantly, the coefficient of variation being near normal conditions for both precipitation and temperature. Nevertheless, the trend line showed (Figure 3A) a slightly increase of temperature and precipitation. The periods from early 1980s to middle of 1990s and from late 2008 to 2013 represent deficit years in temperature while over-years are from 1996 to 2008. On the other hand, over-years in precipitation were observed (Figure 3B) for 1980-1988 and 2008-2015 while deficit in precipitation occurred at the period 1988-2008. So then, the climate conditions fluctuated around three different periods : cool and wet, hot and dry and hot and wet.

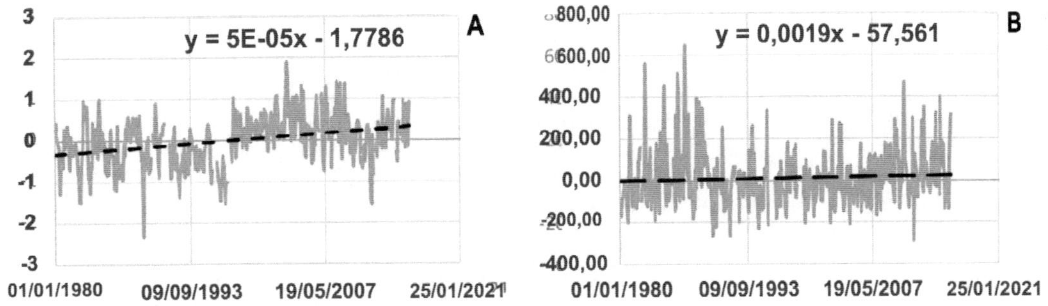

Figure 3: Trend of temperature (A) and precipitation (B) anomalies for 35 years.

3.2. NDVI trend in relation to climatic variables

The regression analysis applied showed that the overall correlation between monthly NDVI and either temperature or precipitation was non-significant. A close to medium positive linear correlation between NDVI and temperature (Coefficient of correlation of 0.47) was found during the long dry season. Yet, only 22% of variation in NDVI can be explained by variation in temperature ($R^2 = 0.22$). A weaker correlation was observed between NDVI and precipitation for the same season (Coefficient of correlation equals to 0.23).

4. Discussion and Conclusion

The system of climatic variables reveals generally stable weather conditions over 35 years. However, the trend analysis revealed a slightly increase of precipitation and temperature even though they are stable, oscillating around normal conditions. The observations of NDVI trend report a similar trend in the vegetation pattern as temperature for the period 2000-2015.

QGIS is a FOSS which has at its disposal functionalities and plug-ins allowing to realize in an easy and fast way tasks which could be tricky to perform otherwise, especially for non advanced users. Indeed, performing tasks as batch processing, for instance, is a convenient option that QGIS provides, espetially while dealing with repetitive tasks over a huge amount of data. R script is a supplementary functionality which opens big possibilities, although it requires some basics knowledge of R language.

The value tool has proved to be particularly useful especially for retrieving time series data. However, it quickly shows its limits when the study area is huge and requires that data are aggregated based on polygons. For this reason, the authors are conducting a similar study in a broader scale (at country level), based on polygon as a spatial analysis procedure to extract MODIS data, in order to obtain further information about the relationship between climate and vegetation in this region and highlight the fact that vegetation may respond differently to climate effects depending on different micro climate.

References

Anonymous, 2017. International research institute for climate and society. earth institute, columbia university. Online, online; accessed May 31, 2017.
URL `http://iridl.ldeo.columbia.edu/dochelp/StatTutorial/Climatologies/index.html#Intro`

Hao, F., Zhang, X., Ouyang, W., Skidmore, A. K., Toxopeus, A. G., 2012. Vegetation ndvi linked to temperature and precipitation in the upper catchments of yellow river. Environmental Modeling Assessment 17(4), 389398.

Mollalo, A., Alimohammadi, A., Shahrisvand, M., Shirzadi, M. R., Malek, M. R., 2014. Spatial and statistical analyses of the relations between vegetation cover and incidence of cutaneous leishmaniasis in an endemic province, northeast of iran. Asian Pacific Journal of Tropical Disease 4(3), 176–180.

NOAA, 2017. How do weather observations become climate data. Online, online; accessed May 31, 2017.
URL `https://www.climate.gov/maps-data/primer/processing-climate-data`

Okin, G. S., Murray, B., Schlesinger, W. H., 2001. Degradation of sandy arid shrubland environments: Observations, process modeling, and management implications. Journal of Arid Environments 47(2), 123144.

Watling, J. I., Brandt, L. A., Mazzotti, F. J., Romanach, S. S., 2013. Use and interpretation of climate envelope models: A practical guide. Online, online; accessed May 31, 2017.
URL `http://crocdoc.ifas.ufl.edu/projects/climateenvelopemodeling/publications/Use%20and%20Interpretation%20of%20Climate%20Envelope%20Models%20-%20A%20Practical%20Guide.pdf`

Facilitate Visualization and Distribution of NASA's Environmental Science Data through Open Standards and Open Source Software for Geospatial

Yaxing Wei[a,*], Zhaoying (Angie) Wei[a], Suresh Vannan[a]

[a]*Environmental Sciences Division (ESD), Oak Ridge National Laboratory (ORNL) P.O. Box 2008, MS 6407, Oak Ridge, TN 37831, USA*

Abstract: This paper introduces the utilization of open standards and open source software for visualization and distribution of geospatial environmental science data at the Oak Ridge National Laboratory Distributed Active Archive Center (ORNL DAAC). The ORNL DAAC is one of the NASA Earth Observing System Data and Information System (EOSDIS) data centers. A big challenge for the ORNL DAAC (https://daac.ornl.gov) is to efficiently manage over a thousand heterogeneous environmental data, collected through field campaigns, aircraft/satellite observations, and model simulations. ORNL DAAC also has to provide tools to easily find, visualize, and access the heterogeneous data. To address this challenge, the ORNL DAAC has leveraged Open Geospatial Consortium (OGC) standards and open source software to develop the Spatial Data Access Tool (SDAT, https://webmap.ornl.gov/ogc). SDAT is a suite of open standards-based web mapping, subsetting, and transformation services and applications that allow users to visualize and download geospatial data in customized spatial/temporal extents, formats, and projections. The open source MapServer/Geospatial Data Abstraction Library (GDAL) powers the backend OGC Web services of SDAT. Open source Javascript libraries, including OpenLayers, GeoExt, and proj4js, were used to create the SDAT Web User Interface and MapWidget, a light-weight Javascript library that allows SDAT visualization to be easily embedded on any webpage. SDAT also provides KML files to enable interactive data visualization in the popular Google Earth application or any KML-compatible client. SDAT provides a common framework and standard service interfaces for ORNL DAAC data holdings. SDAT user interface hides their heterogeneity from end users, and promotes their usage. SDAT facilitates integration of ORNL DAAC data resources with other related data systems. In 2016, SDAT served more than 2 million mapping requests and 72 thousand customized data downloads from over 2500 distinct data users.

*Corresponding author
Email address: `weiy@ornl.gov` (Yaxing Wei)

1. Introduction

National Aeronautics and Space Administration (NASA)'s investments on Earth observation provide a rich set of resources open to a wide variety of users, including Earth and environmental science researchers, decision makers, and public users, to increase our understanding of Earth processes and enhance predictive capabilities that underpin sound decision-making. NASA commits to full and open sharing of Earth science data obtained from NASA Earth observing satellites, sub-orbital platforms, and field campaigns with all users as soon as such data become available (https://science.nasa.gov/earth-science.). Standardization and interoperability are keys to ensure these valuable but heterogeneous data resources can be easily accessed, used, and fused together to address Earth and environmental issues. Free and open source software for geospatial contributed by the community enable data systems and services to be easily created, maintained, and rapidly evolved with relatively low cost. NASA is an active contributor and adopter of such free and open source software (https://software.nasa.gov).

1.1. ORNL DAAC and NASA EOSDIS

The ORNL DAAC for biogeochemical dynamics (https://daac.ornl.gov) is one of NASA Earth Observing System Data and Information System (EOSDIS) data centers managed by the Earth Science Data and Information System (ESDIS) Project. The EOSDIS is a key core capability in NASA's Earth Science Data Systems Program. It provides end-to-end capabilities for managing NASA's Earth science data from various sources satellites, aircrafts, field measurements, and various other programs. The EOSDIS science operations are performed within a distributed system of many interconnected and distributed, discipline-specific DAACs. DAACs are responsible for archive and distribution of Earth science data products. The DAACs serve a large and diverse user community by providing capabilities to search and access science data products and specialized services (https://earthdata.nasa.gov/about). Seamless integration across these distributed nodes is a key to providing consistent service experience to a broad range of users.

The mission of the ORNL DAAC is to assemble, distribute, and provide data services for a comprehensive archive of terrestrial biogeochemistry and ecological dynamics observations and models to facilitate research, education, and decision-making in support of NASA's Earth Science (http://daac.ornl.gov/who.shtml).

1.2. Challenges Faced by the ORNL DAAC and NASA EOSDIS

Within NASA EOSDIS, the ORNL DAAC is unique in terms of the diversity of data it archives, manages, and distributes. ORNL DAAC serves as a repository for data relatively small in size but disparate in variety from principal investigators of field campaigns, airborne missions, remote sensing validation projects, and modeling activities. The ORNL DAAC was formed in 1994 and has archived 1,258 data sets (as of May 26, 2017). These 1000+ data sets contain 373 variables captured/produced from 282 sensors of 125 different sources (satellites, flux towers, airplanes, etc.) by 1410 researchers (Vannan et al. 2016). As illustrated by Vannan et al. 2016, the number of files per data set at the ORNL DAAC ranges from 1 to 92,619 and data volume per data set ranges from 1 KB to more than 10 GB (as of Jan 31, 2017). More importantly, environmental data formats have changed significantly along with the evolution of data capture mechanisms in the past 20 years. Figure 3 of Vannan et al. 2016 shows binary (*.dat) and compressed zip (*.z and*.gz) files constitute 95% of the data files in the ORNL DAAC archive, mainly because binary formats were popular in the 1990s when geospatial file format standards were immature. Data in old binary formats are not readily usable in common data analysis tools and require special data reader programs, often written in Fortran. Total volume of ORNL DAAC data holdings has been increasing more rapidly in recent years and had doubled in 3 years from 90 TB in 2013 to 183 TB in 2015. All these factors raised significant challenges for the ORNL DAAC to serve its user community, including data providers and data users. It is critical to improve the usability of these rich sets of heterogeneous data so that their usage barrier can be further lowered for end users and their value can be maximized and preserved longer into the future.

Expanding the scope from the ORNL DAAC to NASA EOSDIS, data heterogeneity and volume become even bigger challenges. There is an increasing need of improving the interoperability of both data products and data systems so that distributed data archives (DAACs) under the NASA EOSDIS umbrella can be seamlessly integrated to ensure that data providers and users have consistent experience with NASA EOSDIS to archive data they create and find, access, and use NASA Earth science data products.

1.3. Motivation

To address the above described challenges and to improve the usability of increasingly growing ORNL DAAC-archived data products, there is need to:

1. convert data files in outdated formats into current open and commonly-used standard formats
2. enhance data files to be self-descriptive by embedding metadata information following open standards
3. allow users to interactively visualize and explore data prior to download
4. enable users to subset data they need in Geographic Information System (GIS)-ready forms instead of entire original data files
5. improve interoperability of data products and associated visualization and distribution services by following open standards
6. enable the ORNL DAAC data system to be easily integrated into broader data systems

The Spatial Data Access Tool (SDAT) is an effort at the ORNL DAAC targeting the above six goals.

2. OGC Standards

Interoperability is a major goal of the ORNL DAAC SDAT effort. Interoperability between geospatial data systems requires a common set of rules and concepts that define a consistent understanding of the information and operations available in every participating system. Standardization processes and interoperability initiatives such as those of the Open Geospatial Consortium (OGC) try to provide an agreed-on set of such rules and concepts (Vckovski 1999). OGC is an international consortium of companies, government agencies, research organizations, and universities participating in a consensus process to develop publicly available interface specifications. The objective of OGC standards is to promote geospatial information interoperability. OGC Standards adopted by the SDAT, as listed below, provide the foundation for the ORNL DAAC to develop geospatial data visualization and distribution Web services and applications that can be used by end data users and common software and tools that support OGC standards.

- Web Coverage Service (WCS)
- Web Feature Service (WFS)
- Web Map Service (WMS)
- Keyhole Markup Language (KML)
- Styled Layer Descriptor (SLD)
- Well-Known Text Coordinate Reference System (WKT CRS)

WCS supports electronic interchange of geospatial data as coverages that is, digital geospatial information representing space-varying phenomena that relate a spatio-temporal domain to a range of properties. A WCS provides access to coverage data in forms that are useful for client-side rendering, as input into scientific models, and for other clients (OGC 2005). At ORNL DAAC, the WCS standard is used to distribute geospatial imagery data files like Land Use/Land Coverage Change data and ecosystem model simulation data (Wei et al. 2009).

WFS provides an interface allowing access to discrete geospatial feature data encoded in Geography Markup Language (GML) (OGC 2007b). Unlike coverages, geospatial features are described by a set of properties where each property can be thought of as a name, type, value tuple (OGC 2010). The vector data files archived at ORNL DAAC such as field campaign site location information, Global River Discharge field data fall into this category and can leverage WFS for data distribution.

WMS, unlike WCS or WFS that provides access to geospatial data itself, produces styled maps of georeferenced data, including both coverage and feature data. A map is defined as a visual representation of geospatial data (OGC, 2001). WMS can be used to visualize data prior to downloading. SLD defines an encoding that extends the WMS standard to allow user-defined symbolization and coloring of geographic feature and coverage data (OGC 2007a). SDAT utilizes SLD to allow users to easily provide their own or import existing map styles when making WMS requests.

KML provides an XML notation for expressing geographic annotation and visualization within Internet-based, two-dimensional maps and three-dimensional Earth browsers, e.g. Google Earth and NASA World Wind (OGC 2015b).

WKT CRS is an OGC standard jointly developed by International Organization for Standardization (ISO) and OGC. It provides well-known text string representation of coordinate reference systems that follows the provisions of ISO 19111:2007 and ISO 19111-2:2009 (OGC 2015a). WKT CRS is leveraged by SDAT to provide standard representation of the CRSes associated with geospatial data and maps generated by SDAT, so that they can be automatically recognized and imported into majority of commercial and open source GIS software.

Among those OGC standards supported by SDAT, WMS and KML have been adopted by NASA ESDIS Standards Office (ESO) and recommended for use in NASA Earth Science Data Systems.

3. Spatial Data Access Tool

3.1. Overview

SDAT is a Web-based framework developed at the ORNL DAAC to ensure archived geospatial data are in open and standard formats. It provides easy and interoperable OGC Web services for users and compatible clients to retrieve geospatial data and maps based on their specific needs. The SDAT framework as illustrated by Fig. 1 contains a family of Web-based services and applications for geospatial data visualization and exploration, transformation (i.e. subset, reformat, resampling, and reprojection), and data download.

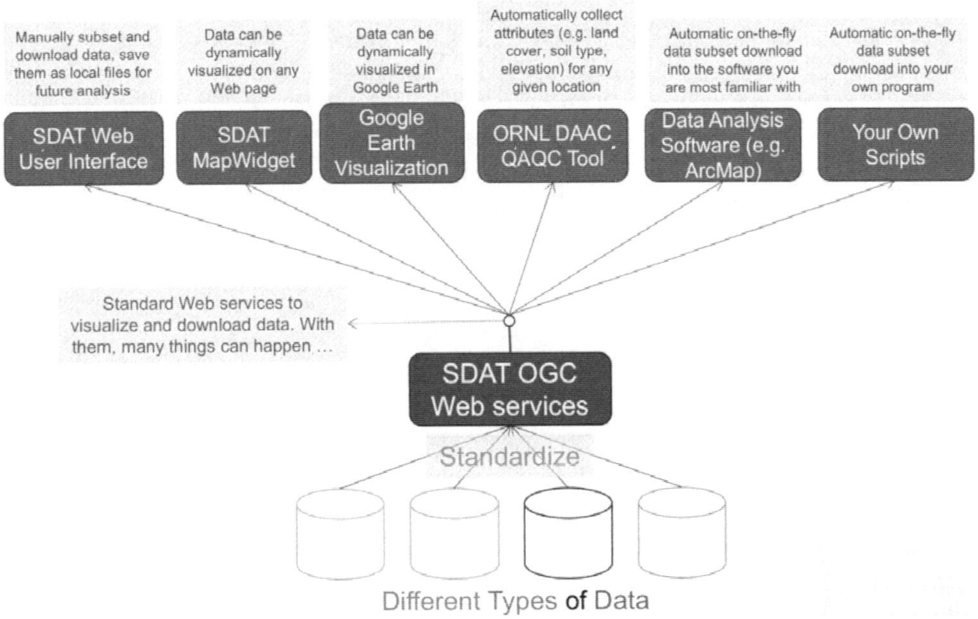

Figure 1: Spatial Data Access Tool (SDAT) framework

At the foundation of SDAT are well-curated geospatial data in open standard formats, which are then ingested into SDAT OGC Web services (i.e. WCS, WFS, and WMS) and made available to users through standard service interfaces. The SDAT Web User Interface (UI), the SDAT MapWidget, and Google Earth Visualization are the three major Web applications in the SDAT framework that provide advanced visualization, transformation, and data download capabilities to users who are not familiar with OGC standards. Interoperability achieved through SDAT OGC Web services allows other applications, such as the ORNL DAAC Identify Tool (https://webmap.ornl.gov/qaqc/identify), to be easily developed utilizing SDAT's OGC Web services. Of course, many data analysis software, including common commercial and open source GIS tools, already supports OGC standards and can consume SDAT OGC Web services as remote geospatial data resources directly inside analysis workspace.

At this moment, SDAT WFS is in testing phase. The ORNL DAAC is investigating the benefits that WFS can bring to users and the amount of effort needed to curate feature-based geospatial data, such as field and airborne observations, and make them available through WFS. As a data center focusing on field campaigns and land validation, OGC WFS may have a lot of potentials for the ORNL DAAC and its users.

3.2. SDAT Implementation

3.2.1. Data Curation

The SDAT workflow starts with curating a wide variety of geospatial data into open standard formats. GeoTIFF is the most popularly-used open format for geospatial imagery data in the

GIS community. It was selected as the primary format for imagery data prepared for SDAT WCS and WMS. When SDAT first started, it focused on data sets that had been archived at the ORNL DAAC in the past. Important data sets (e.g. most downloaded), especially those stored in outdated formats, were prioritized for the SDAT effort. Data files in pure binary, text, or other legacy formats were converted into the GeoTIFF format and enhanced with rich metadata information, including standard CRS representation, variable names, and data units. In recent years, with the maturing of the Climate Forecast (CF) convention (Eaton et al. 2011), the NetCDF format is gaining popularity rapidly. Well-curated, especially spatially and temporally, CF-compatible netCDF data format is another option for preparing imagery data for SDAT. Recently, all SDAT data preparation requirements have been incorporated into ORNL DAAC's standard data publication workflow. This ensures every geospatial data set prepared by the ORNL DAAC's data publication team is ready for SDAT without any further processing, which helped SDAT to shift its focus from past data sets to present and future ones.

As of September 2017, 147 ORNL DAAC data sets (containing 8,185 data files) have been curated and added to the SDAT and OGC Web services.

3.2.2. Metadata Database

Implementation of SDAT relies on a centralized metadata database, which stores metadata information for both data sets (data set table) and data files (data granule table), to automate the configuration of data layers and the mapping between layer identifiers and data files. Data layer is a concept used in OGC Web services. In SDAT, a data set contains multiple data files, or granules, and each data granule is further mapped onto one or more data layers, depending on the number of variables a data file contains.

The data set table contains field dataset_id, which stores the unique ID for each data set. Data set IDs are consistent with ORNL DAAC's core database and provide a logical linkage between SDAT and the ORNL DAAC data system. Other general metadata, such as dataset name, description, citation, and keywords, of the data set table are also synchronized with the ORNL DAAC's core database. To automate the generation of a common color scheme for all layers in a dataset, dataset-level data value statistics, i.e. maximum, minimum, mean, and standard deviation values, are included in the data set table.

In the data granule table, the granule_id field, specifically created for SDAT, contains the unique identifier of each data granule, which is used to construct data layer descriptors in the SDAT OGC Web services. Granule-level metadata contain descriptive information (filename, variables, data type, data unit, resolution etc.), file-level statistical information (maximum, minimum, mean, standard deviation, and missing values), as well as the spatial and temporal extent information (i.e. bounding box, native projection, start time, end time, time series). The statistical metadata is used to automatically generate a color scheme for each individual data layer in OGC Web services. The spatial and temporal extent metadata facilitate the subsetting based on geographical/temporal extent of the data.

3.2.3. SDAT OGC Web Services

The core components of SDAT are the OGC Web services (i.e. WCS, WFS, and WMS) that provide geospatial data visualization, transformation, and access capabilities. To maximize the usage of these Web Services, we chose to support the most widely-used versions of OGC standards, which are not necessarily the latest versions. WCS version 1.0.0, WMS version 1.1.1, and WFS version 2.0 are currently supported in SDAT.

Implementation of the SDAT OGC Web services is based on MapServer, which is an Open Source platform for publishing spatial data and interactive mapping applications to the web. Originally developed in the mid-1990's at the University of Minnesota, MapServer is released under an MIT-style license (http://mapserver.org). When the SDAT effort started, there were multiple mapping platform options to choose from, including the commercial ArcIMS/ArcGIS Server and the open source MapServer and GeoServer. Commercial platforms were not preferred since a non-open platform will significantly increase development and maintenance cost in the long run. MapServer was eventually selected as the mapping engine for SDAT for the following reasons: 1) MapServer had outstanding geospatial mapping performance, especially with FastCGI enabled on Linux operating systems (OSGeo 2011); 2) It supported a large array of imagery and feature data formats using the open source GDAL/OGR library (http://www.gdal.org); and 3) Even though MapServer lacked a nice Web Graphic UI for managing data layers, its mapfile-based configuration mechanism allowed data layers to be easily managed in MapServer and automated through shell scripts or MapScript API.

MapServer, along with the open source GDAL/OGR and PROJ libraries, provide the geospatial data mapping and transformation capabilities of the SDAT OGC WMS and WCS. The SDAT WCS supports spatial and temporal subsetting. 6 different popular data formats (ArcInfo/ASCII Grid, Erdas Imagine, GeoTIFF, NetCDF, NITF, and XYZ) are supported in the SDAT WCS as output formats for users/clients to choose from. For each data format, users/clients can choose what internal data type (floating point, integer, or byte) to use. These 6 data formats include both open and proprietary formats to better support user needs. At this moment, each data layer available through SDAT OGC WMS and WCS supports a fixed list of target CRSes, which include the data layer's native CRS, commonly-used CRSes (e.g. EPSG:4326 and Web Mercator) across all data layers, and project-specific CRSes (for example, the ABoVE project (https://above.nasa.gov) has defined CRSes suitable for data sets produced by ABoVE). In the future, SDAT OGC Web services will be enhanced to allow users/clients to provide customized CRSes instead of choosing from a list of pre-defined CRSes.

3.2.4. SDAT Manager: automate data configuration into SDAT

Making well-curated data available through SDAT OGC Web services with low effort is one major design goal of SDAT. To minimize human effort needed, the SDAT Manager was developed to automate majority housekeeping tasks. The SDAT Manager is a Web-based dashboard application accessible only by SDAT team members. Supported by backend server-side scripts and the SDAT metadata database, the SDAT Manager provides team members easy controls for automatic creation of map color schemes and mapfiles needed by MapServer. The Manager also supports the generation of static preview images for each data layer, update of OGC Capabilities XML files when data layers are added or updated, creation of KMZ packages for Google Earth visualization, production of SLD files containing default color schemes that users can customize for their own needs, and miscellaneous metadata database management tasks (Fig. 2). The SDAT Manager has significantly streamlined the data publication process for SDAT and its OGC Web services.

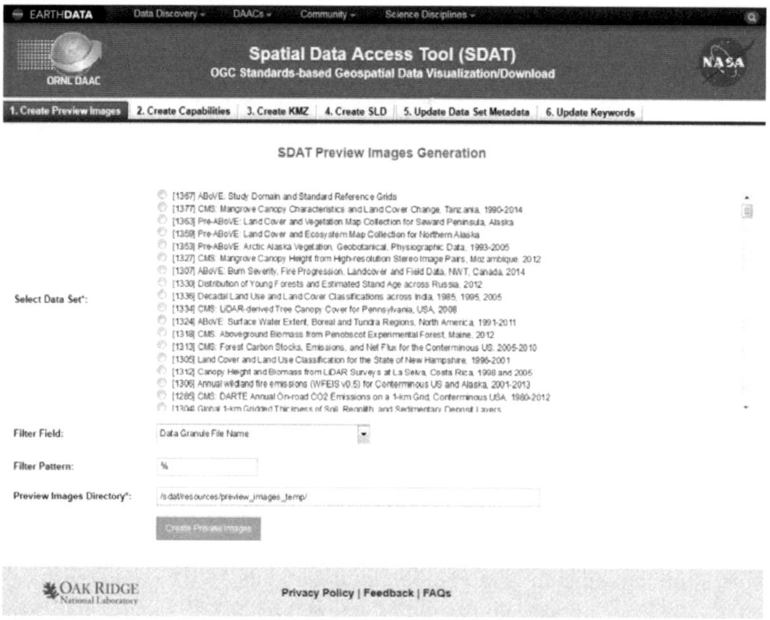

Figure 2: SDAT Manager dashboard

3.2.5. SDAT Key Applications and Use Scenarios

3.2.5.1. SDAT Web UI. To provide advanced capabilities of SDAT OGC Web services to non-expert users who are not familiar with OGC standards, the SDAT Web UI (https://webmap.ornl.gov) was developed to allow users to search, visualize, explore, and download transformed data they need through an intuitive Web user interface. The SDAT Web UI organizes data in two levels: data set level and data granule level. A data set contains one or more data granules (or files), following the same data organization as the ORNL DAAC data archive.

As shown in Fig. 3, the home page of SDAT Web UI presents users a paginated list of data sets. Users can filter data sets through two groups of filters: 1) science disciplines, which are topic-level keywords in Earth Science category of Global Change Master Directory (GCMD); and project names. These 167 data sets in SDAT are categorized into 8 science disciplines and 12 projects. The home page also offers keyword-based search functionality, which can be combined with the science disciplines and project filters to further refine search results. For example, Fig. 3 displays 12 matched data sets (out of 167) after applying the Biosphere science discipline filter, the LBA project filter, as well as keyword land. Order of data sets in the search results is based on when they were published into SDAT and most recently added data sets are displayed on top.

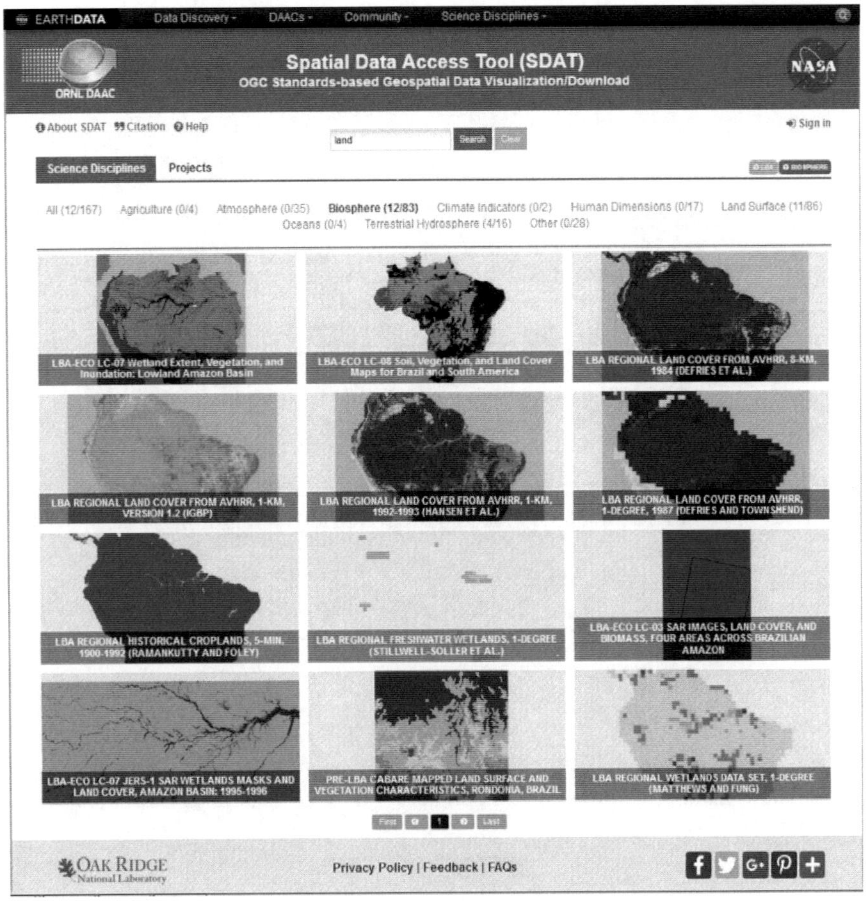

Figure 3: Home page of SDAT Web UI: List of data sets with filters applied

The data download page (Fig. 4) contains 3 sections. The first section displays granule-level metadata, including spatial extent and available time steps. The second section contains a map viewer, dynamically generated by the SDAT OGC WMS service, for users to interactively visualize the geospatial data. The third section gives users control to customize geospatial data download by setting WCS (or WFS) parameters, which provide a series of data transformation options, such as subsetting, reformatting, resampling, and reprojection.

Data transformation options offered by SDAT allow users to download data in ready-to-use forms. For example, to analyze surface water extent changes in boreal and tundra regions, users can compare data of ABoVE surface water extent in two adjacent decades: 2000s and 2010s (Carroll et al. 2016) (https://webmap.ornl.gov/ogc/dataset.jsp?ds_id=1324). They first choose the CRS (e.g. Web Mercator) needed. Then to specify subset region, users can draw a bounding box over the map, say an area surrounding the Beaver Hill lake in Canada, or enter the spatial extent manually in the Spatial Extent boxes. After selecting the output format and other transformation parameters, users can request data transformation through SDAT WCS and download data with a single click, once signing in via Earthdata login (https://urs.earthdata.nasa.gov). Data downloaded through SDAT are in a ready-to-use data format and can be directly imported into data analysis software, such as QGIS (Fig. 5), to calculate differences between surface water extents in the 2000s and 2010s in the Beaver Hill lake area. Areas in red color represent shrinking of surface water extent during the last two decades. SDAT facilitates such exploratory data visualization and analysis through standards-based data delivery and visualization.

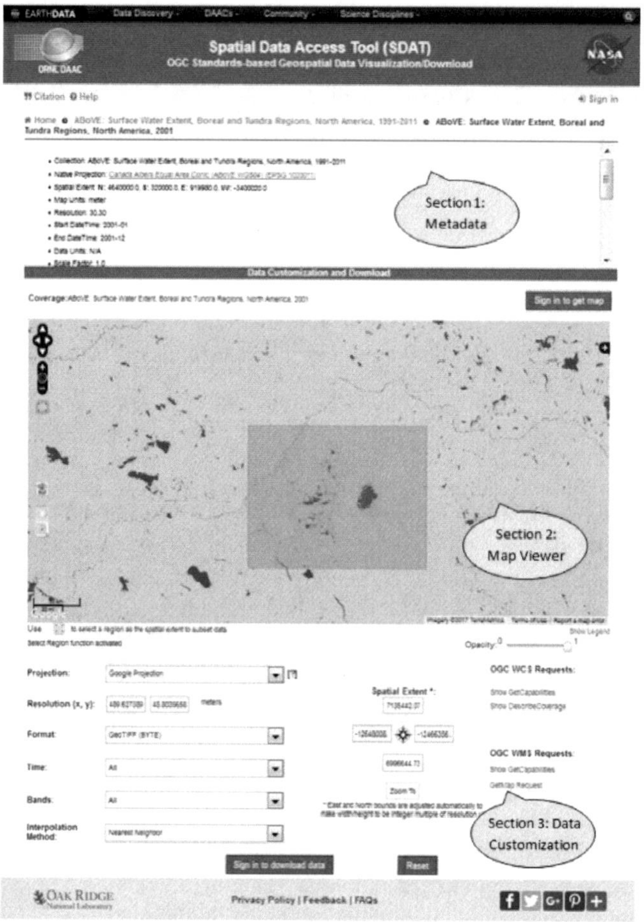

Figure 4: SDAT Web UI - data visualization and download page for ABoVE surface water extent in the 2000s

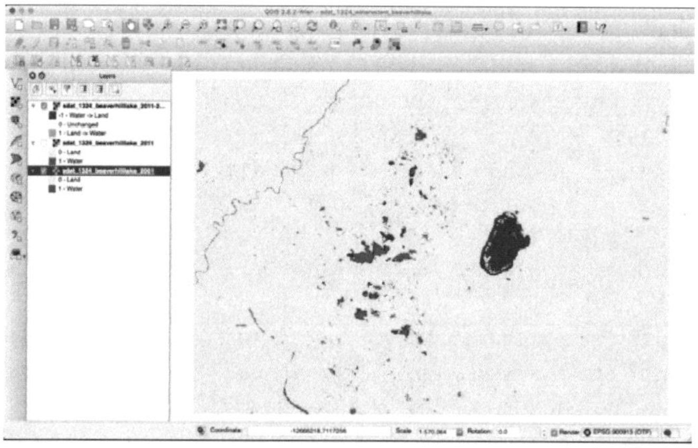

Figure 5: Analysis of surface water extent changes in the Beaver Hill Lake area in QGIS

Implementation of the SDAT Web UI utilizes open source software and libraries, including JQuery, OpenLayers, and Proj4js.

3.2.5.2. SDAT MapWidget. SDAT MapWidget is a JavaScript library that allows a light-weight interactive map viewer to be easily embedded onto any Web page to provide live data visualization, very similar to how Google Map viewer works. SDAT MapWidget is built on top of the open source OpenLayers and GeoExt libraries. The ORNL DAAC has incorporated SDAT MapWidget as a standard component on its data set landing pages to allow users interactively to visualize and explore data without leaving landing pages. This helps users to better evaluate the fitness-to-use of geospatial data through deeper, intuitive, and closer interaction. Fig. 6 shows the SDAT MapWidget viewer on the landing page (https://doi.org/10.3334/ORNLDAAC/1290) of data set NACP NAFD Project: Forest Disturbance History from Landsat, 1986-2010 (Goward et al. 2015).

Figure 6: SDAT MapWidget viewer for NACP NAFD Project: Forest Disturbance History from Landsat, 1986-2010

3.2.5.3. Google Earth Visualization. For each data set, SDAT provides a KML/KMZ file that can be dynamically visualized in KML-compatible clients, such as Google Earth. We leveraged an XSLT WMS to KML converter (Benedict and Sanchez-Silva, https://github.com/rnz0/wms2kml) and enhanced it to create KML/KMZ files from WMS Capabilities documents. These KML/KMZ files are configured to retrieve new maps on-the-fly from SDAT OGC WMS service whenever Google Earth viewport changes. Google Earth provides nice features and resources, for example, time-series animation, high-resolution background imageries, and historical Landsat imageries, that enhance users' data exploration experience. Fig. 7 presents visualization of DARTE Annual On-road CO2 Emissions on a 1-km Grid (Gately et al. 2015) in Google Earth. The time series animation in Google Earth allows users to easily explore changes of annual CO2 emissions from on-road transportation in the Knoxville, TN area from 1980 to 2012. Comparing 3 maps in Fig. 7, on-road CO2 emissions in 1990 were lower than in 1980 and 2000, which could be possibly affected by the early 1990s economic recession.

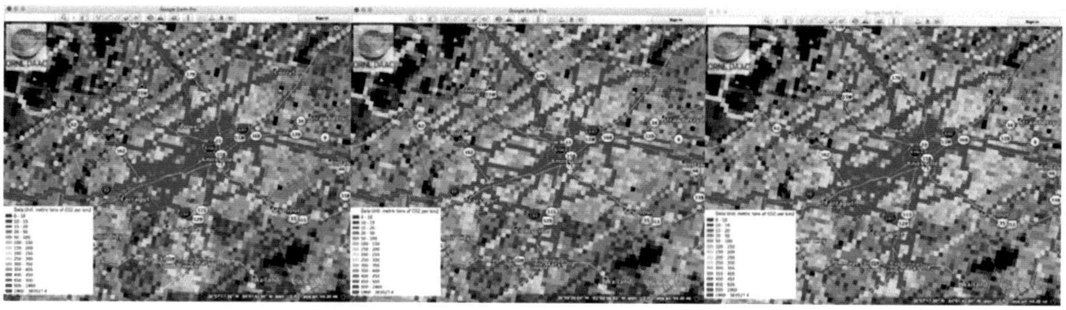

Figure 7: Visualization of annual on-road CO2 emissions in Knoxville, TN area in 1980, 1990, and 2000 in Google Earth

3.2.6. Use SDAT/OGC Data in Common GIS software and programs

For advance users who have knowledge of OGC standards, they can directly access SDAT OGC Web services from many popular GIS software (e.g. ESRI ArcMap and QGIS) and their own programs (e.g. Matlab and Python). In this way, there is no need to download data onto local machines. Remote data available through SDAT OGC Web services can be consumed directly in these software and programs as if they exist locally.

3.3. SDAT Citation

Citation of data products, especially those dynamically transformed through Web services, is being actively discussed within NASA ESDIS and broader data communities (Hausman et al. 2017) (Smith et al. 2016) (Rauber et al. 2015). The ORNL DAAC is a pioneer on providing citations for not only data products but also data services it offers. In the past, data users cited data downloaded via SDAT through footnote or acknowledgement. In early 2017, the ORNL DAAC registered a DOI and created a formal citation for SDAT.

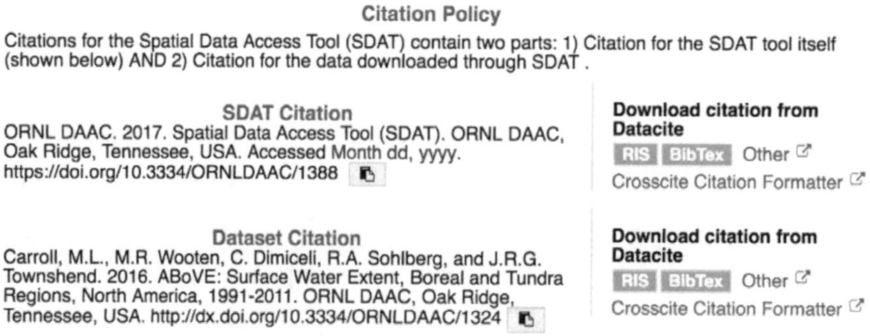

Figure 8: SDAT citation policy popup for the ABoVE Surface Water Extent data set

An initial SDAT citation policy has been set up and provided to users on the SDAT Web UI. As shown in Fig. 8, the SDAT citation policy suggests including citations for SDAT itself and the data set downloaded. SDAT also leverages DataCite and CrossCite Citation Formatters to provide DOI citations in other formats.

4. Integration of SDAT/OGC with NASA EOSDIS and Broader Data Systems

4.1. SDAT/OGC - NASA EOSDIS

NASA EOSDIS provides an enterprise data infrastructure that can seamlessly integrate the distributed data systems of all data processing systems and DAACs. This integration involves many different aspects. Metadata integration is achieved through the Unified Metadata Model (UMM) (https://earthdata.nasa.gov/about/science-system-description/eosdis-components/common-metadata-repository/unified-metadata-model-umm) and the Common Metadata Repository (CMR) (https://earthdata.nasa.gov/about/science-system-description/eosdis-components/common-metadata-repository). The SDAT metadata database will be fully integrated with the ORNL DAAC core database in the future so that SDAT metadata elements will be incorporated into UMM and managed by the CMR. In return, SDAT Web UI will be able to leverage the Earthdata Search API to provide data search capability, on both data set and file levels. Efficient data access and transformation is key to utilization of EOSDIS data. The CMR and the Earthdata Search application (https://search.earthdata.nasa.gov) are capable of brokering data orders to avoid ordering applications or users needing to know DAAC specific ordering systems [24]. This brokering capability has been enabled for OGC WMS services to dynamically display maps on the search result map viewer. NASA EOSDIS is actively seeking solutions to add brokering capability to CMR and Earthdata Search application for OGC WCS services hosted at each individual DAACs. Data in SDAT are freely available to any user, but require user identification per NASA policy to help track detailed data usage at the individual user level, enabling both accurate metrics on data usage and the ability to send critical data update messages (NASA 2016). SDAT utilizes NASA EOSDIS Earthdata Login for user registration, authentication, and identification and provides single sign-on support across NASA EOSDIS.

4.2. SDAT/OGC - GEOSS

Interoperability enabled by SDAT OGC Web services allows ORNL DAAC-archived data to be easily integrated into broader data systems, such as the Global Earth Observation System of Systems (GEOSS), without any additional effort on the ORNL DAAC side. This is a huge benefit of OGC standards and maximizes the usage and value of ORNL DAAC data. GEOSS is a set of coordinated, independent Earth observation, information and processing systems that interact and provide access to diverse information for a broad range of users in both public and private sectors. GEOSS adopted a Service-Oriented Architecture (SOA) and set interoperability as its foundation (Khalsa et al. 2009). The GEOSS Portal offers a single Internet access point for users seeking data, imagery and analytical software packages relevant to all parts of the globe. The GEOSS Portal has integrated all data layers of SDAT WMS through dynamically pulling and parsing Capabilities XML of SDAT WMS. These SDAT WMS layers are available for users to search and visualize directly within the GEOSS Portal (http://www.geoportal.org).

5. SDAT/OGC Usage Analysis and Future Improvements

SDAT has been gaining popularity among a broad range of users in the past few years. One reason is the improvements, on the aspects of both performance and usability, we have been continuously making to SDAT. The other reason is the increased support of OGC standards in common GIS and data analysis tools, such as ArcMap, QGIS, Matlab, and R, that people use on a daily basis.

As shown in Fig. 9, usage (number of data downloads, volume of data downloaded, and number of unique user IP addresses) of SDAT WCS has been consistently increasing in the past 6 years (2011-2016). Number of unique WMS user IP addresses has also been consistently increasing in 2011-2016. But the usage of SDAT WMS had a significant jump in 2014 and remained high usage rate thereafter. This was possibly caused by increased visibility of SDAT WMS through integration into broader data systems, which attracted some power geospatial mapping users, especially those based on Cloud computing platforms.

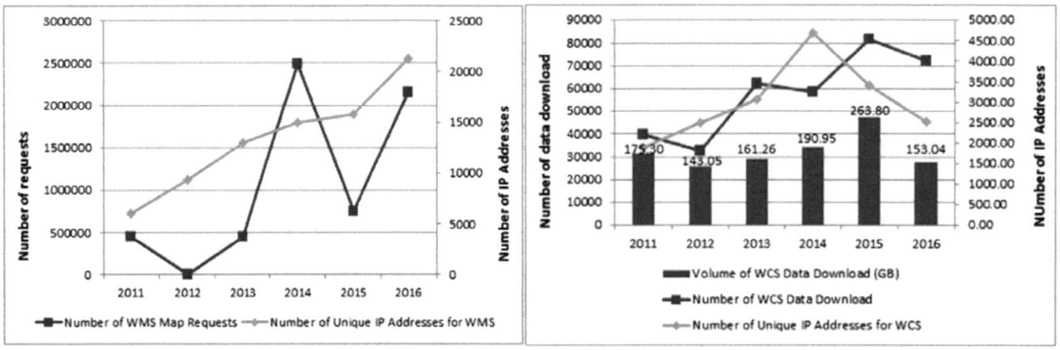

Figure 9: SDAT WCS/WMS usage statistics

We performed further analysis on the 72,000 SDAT WCS data downloads in 2016 to discover to what extent users had utilized the transformation capabilities provided by SDAT WCS. Table 1. shows that GeoTIFF, Arc/Info ASCII Grid, and NetCDF were the most popular data formats users selected. Majority users (85.1%) leveraged the subsetting capability. Close to half (43.9%) users chose to request data in different spatial resolutions. But only 3.7% users chose to reproject data onto a CRS different from their native CRSes. Reprojection is a critical issue to fuse heterogeneous geospatial data from different sources. Even though SDAT WCS provides reprojection capability, but it was not fully utilized. The main reason is that only limited and fixed target CRS options are available for users. We will enhance SDAT OGC Web services to allow users/clients to provide customized CRSes to request data and maps in the future.

Arc/Info ASCII Grid	GeoTIFF	Erdas Imagine	NetCDF	NITF	XYZ
12.4%	77.8%	0.65%	7.1%	0.14%	1.91%

Table 1: Statistics on requested data formats of SDAT WCS service

One most requested feature that will be added to SDAT in the future is the SDAT Workspace for the SDAT Web UI. The SDAT Workspace is aimed to solve multiple issues. It will allow users to select multiple data files from multiple data sets and add them onto a common viewer for visualization. Also, users can specify a single set of data transformation criteria, e.g. target CRS, spatial extent/resolution, temporal range, and data format, and use them to batch download multiple data files through SDAT WCS with one single click. Results data files, along with metadata and transformation provenance information, will be packed together and accessible by users.

OGC WFS will be further investigated and fully incorporated into the SDAT framework to provide visualization and download capabilities for feature-based geospatial data products, including observations made at field locations and along aircraft flying trajectories.

6. Conclusion

NASA's investments on Earth observation provide a rich set of valuable resources open to a wide variety of users, including Earth and environmental science researchers, decision makers, and public users, to increase our understanding of Earth processes and enhances predictive capabilities that underpin sound decision-making. NASA EOSDIS provides a distributed infrastructure that links distributed DAACs, including the ORNL DAAC, through interoperable interfaces and allows a wide variety of users to easily find, visualize, explore, and access Earth science data resources. The open standards and open source software-based SDAT increases usability of the environmental science data products archived at the ORNL DAAC by assuring data are stored in open and self-descriptive standard formats, enabling users to interactively visualize and explore data prior to downloading, allowing users to download data they need in GIS-ready format instead of entire original data files. More importantly, SDAT and the OGC Web services improve the interoperability of the environmental science data, enable the ORNL DAAC data system to be easily integrated into broader data systems, and maximize the usage and value of NASA's environmental science data. The ORNL DAAC SDAT has gained great popularity in the past few years. But there are still many aspects needing improvement, including dynamic data citation, improved data transformation capabilities, better integration with the NASA EOSDIS infrastructure, and the SDAT Workspace concept that allows users to overlay multiple data layers for visualization and batch download multiple data with the same set of transformation criteria.

References

Carroll, M. L., Wooten, M. R., Dimiceli, C., Sohlberg, R. A., Townshend, J. R. G., 2016. Surface water extent, boreal and tundra regions, north america, 1991-2011. ornl daac, oak ridge, tennessee, usa. Online, online; accessed May 31, 2017.
URL `https://doi.org/10.3334/ORNLDAAC/1324`

Eaton, B., Gregory, J., Drach, B., Taylor, K., Hankin, S., Caron, J., Signell, R., Bentley, P., Rappa, G., Hock, H., Pamment, A., Juckes, M., 2011. Netcdf climate and forecast (cf) metadata conventions. Conventions 1.6.

Gately, C. K., Hutyra, L. R., Wing, I. S., 2015. Cms: Darte annual on-road co2 emissions on a 1-km grid, conterminous usa, 1980-2012. ORNL DAAC, Oak Ridge, Tennessee, USA.

Goward, S. N., Huang, C., Zhao, F., Schleeweis, K., Rishmawi, K., Lindsey, M., Dungan, J. L., Michaelis, A., 2015. Nacp nafd project: Forest disturbance history from landsat, 1986-2010. ORNL DAAC, Oak Ridge, Tennessee, USA.

Hausman, J., Stall, S., Gallagher, J., 2017. Recommendations of nasa esdswg 2016-2017 software citations working group. Citations Working Group.

Khalsa, S. J. S., Nativi, S., Geller, G. N., 2009. The geoss interoperability process pilot project (ip3). IEEE Transactions on Geoscience and Remote Sensing 47(1), 80–91.

NASA, 2016. Esdis project. EOSDIS Handbook 1.

OGC, 2005. Opengis web coverage service (wcs) implementation specification. OGC 05-076 1.0.0.

OGC, 2007a. Opengis geography markup language (gml) encoding standard. OGC 07-036 3.2.1.

OGC, 2007b. Styled layer descriptor profile of the web map service implementation specification. OGC 05-078r4 1.1.0.

OGC, 2010. Opengis web feature service 2.0 interface standard. OGC 09-025r1 2.0.0.

OGC, 2015a. Geographic information - well-known text representation of coordinate reference systems. OGC 12-063r5.

OGC, 2015b. Ogc kml 2.3. OGC 12-007r2.

OSGeo, 2011. Wms performance shootout 2011. Online, online; accessed May 31, 2017.
URL `https://www.slideshare.net/gatewaygeomatics.com/wms-performance-shootout-2011`

Rauber, A., Proll, S., Citation, W. R. D., 2015. Scalable dynamic data citation approaches, reference architectures and applications. Draft Version.

Smith, A. M., Katz, D. S., Niemeyer, K. E., 2016. Force11 software citation working group. software citation principles. PeerJ Computer Science 2:e86.

Vannan, S., Beaty, T. W., Cook, R. B., Wright, D. M., Devarakonda, R., Wei, Y., Hook, L. A., McMurry, B. F., 2016. A semi-automated workflow solution for data set publication. ISPRS Int. J. Geo-Inf. 5(30).

Vckovski, A., 1999. Interoperability and spatial information theory. In Interoperating geographic information systems 495, 3137.

Wei, Y., Santhana-Vannan, S. K., Cook, R. B., 2009. Discover, visualize, and deliver geospatial data through ogc standards-based webgis system. 17th International Conference on Geoinformatics, Fairfax, VA, 1–6.

GEOPOLL - Integrate Cartographic Questions in Web Forms, Polls or Surveys

Adrien Bigler[a,*], Olivier Ertz[a], Daniel Rappo[a], Sarah Composto[b], Florent Joerin[b], Maude Luggen Risse[b]

[a] *HEIG-VD / Media Engineering Institute, University of Applied Sciences Western Switzerland. E-mail: mei@heig-vd.ch*

[b] *HEIG-VD / INSIT, University of Applied Sciences Western Switzerland. E-mail: insit@heig-vd.ch*

Abstract: Most of the web forms, polls or surveys are composed with classical input fields (check boxes, radio buttons, select lists, etc.) and a lot of standards web forms builders or services (e.g. Typeform, WuFoo, Google Forms, Survey Monkey, etc.) help to build and deploy them. Nowadays, offering a text area to catch a ZIP code or a select list to point out a country remains the best way to explicitly collect geospatial data. But what about mapping interfaces to integrate cartographic questions and/or cartographic answers as a more suitable solution? How to let form's respondents indicate in which area(s) they engage in certain activities? The only way to solve these issues is by coding it yourself. This paper shows the design and technical problematic and try to give some solutions to build an efficient UX for the respondents and a standard answers treatment for the data analysts. This paper will also highlight the capacities of open source JavaScript library like OpenLayers to build the solution. Our interface was tested by the cities of Lausanne and Pully (Switzerland) and thousands of people answered to these cartographic forms on mobile and desktop. We would like to share some good practice and lessons learned.

*Corresponding author

Email address: `adrien.bigler@heig-vd.ch` (Adrien Bigler)

1. Introduction

This article reflects the development of two surveys made for the municipalities of Lausanne and Pully in Switzerland. Indeed, the planning department and the political representatives of these two cities in the French-speaking part of Switzerland have launched some initiatives whose aim is to involve citizens in the decision-making processes related to future developments of their territories.

In one case, the city of Pully wants to give a new face to its city center (Pully 2017). Previous experiences have proven that the rearrangement of a city center is a sensitive subject, this is why, late 2015, the municipality decided to conceive a public consultation. The target audience for this consultation were the inhabitants, the workers and the other users of the city center. In the second case, starting from 2015, the city of Lausanne is renewing its train station as the travelers flow will double by 2030. This will impact the surroundings and their users. The city wanted to make a public consultation were the travelers, the workers and the neighborhood residents can participate to rethink the train station surroundings (Lausanne 2017). For both cases, people were involved in multiple activities like workshops, conferences, debates, paper and web surveys.

2. Problem Definition

In urban planning, what is commonly referred to as a "participatory initiative" intensively uses tools or methods of implementation of many kinds: public consultation, seminar, workshop, interview, exhibition, etc. Even Pully and Lausanne municipalities are already used to publish various online surveys with the dedicated tools or providers available on the market (e.g. WuFoo, Google Forms, Type Forms, JotForm, Survey Monkey, etc.) which have greatly contributed to democratize this type of actions. However, given that issues related to urban planning are closely related to the people and activities in spaces of various scales, they are usually not easy to grasp as soon as we use the traditional input types that make up web survey forms (radio buttons, check boxes, lists, sliders, text, etc).

It is then obvious that techniques from user-generated content have to be considered, especially from VGI (Volunteered Geographic Information). As an example, when it is about a citizen science project (Haklay 2013) which requires contributors to collect data about biodiversity (Ingensand et al. 2015), a cartographic user interface is essential. In other words, the map is always at the heart of the solution. This is also true for the projects with Pully and Lausanne municipalities and the kind of surveys that was required. The literature shows that many design aspects of web forms have been explored of which Bargas-Avila et al. 2010 does classify the design in five topics: form content, form layout, input types, error handling, form submission. If guidelines to improve efficiency, effectiveness and user satisfaction have been defined for classical input types like checkboxes, radio buttons or drop-down menus, the case of using cartographic user interface as another input type has never been investigated. In a way, both demands coming from the municipalities were the opportunity to consider them as case studies to push forward the design process with the map "just" as a new generic input type of a form. It is to introduce a new type of question in a survey, the cartographic question that involves an interaction with a map. In the following, we describe some essential aspects of the design process from which we extract some guidelines.

Also, the following requirements and technical specifications had to be considered:

• surveys developed for the web, in other words, no native source code to manage;
• responsive design so as to reach and satisfy the greatest number of potential users;
• single-page mode, so that respondents can immediately evaluate how long is the task to fill in the survey.

3. Cartographic Questions?

When it comes to geographic data in a web form, we can think of postal address input fields. Usually, they are text fields or drop-down lists (for example, to choose a country). Using predefined directories of the possible values, auto-completion can make the task of answering the question easier or more efficient from the user's point of view. The more advanced interfaces provide at most a way to indicate an address by clicking on a map and using reverse-geocoding to get the underlying address, but this type of service is by far not very widespread.

In our case, respondents are asked to answer questions about their activities inside a urban space and how they perceive this space. For example:

• Where do you live? Where do you work?
• What use do you have of this area?
• In your opinion, where is the city center?
• Etc.

Some of the expected answers have a perceptual nature, they are all strongly related to map areas the respondents are aware of and should recognize because they often visit them. Therefore it seems obvious that the use of a map is more apt to collect such expected answers. But what are the issues to consider to design a cartographic question ?

First of all, it is important to think about the content of the expected answers according to the different points of view, like:

• the analyst: what is the level of precision expected, for what use or for what analysis?
• the respondent: who would agree to give the precise address of her/his employer, or her/his exact home address? How long will it take to respond? Who is smart enough to define the rigorous perimeter of a city center? Can everyone read a map?

Then, in term of interaction design, even if the collected geographic data are most often at the neighborhood level and because the territory concerned may be relatively wide (at the scale of a city), the interface must certainly offer the standard functionalities that allow navigating a map (zoom, pan).

And for the main issue, that is about how to collect the answers, two approaches are worth to be considered: input by drawing and input by selection.

3.1. Input by drawing

For this approach, the answer is a "portion of territory" which is supposed to be digitized by the user. After an analysis of all the cartographic questions related to our two case studies, it shows that four types of spatial entities may be possible answers, that is with: a point, a circle, a polygon area, or a scrawled area (drawn freehand). Indeed, none of the considered questions could expect a linear object as a response.

The approach by drawing manually the response is interesting as it gives the greatest freedom to the users. The design is objective without any preconceived vision, it delegates to the respondent the expression of the subjective part of her/his vision.

Adding a point marker is the simplest operation (Figure 1a). On the other hand, it suppose from the user to be precise, which is often not advisable considering the above mentioned respondent point of view: as soon as it is related to privacy, the respondent may hesitate to be precise enough, and we do not want to force the user to "lie" about the location.

Using a circle allows to remain vague with an approximation of the location (Figure 1b). However, the average user is not used to draw circles on a map and we aim for surveys easy to pick up for a wide audience.

The same goes for drawing a polygon area (Figure 1c), it is even worse. User testing has shown how some users are unable to close the polygon. Furthermore, often users waste unnecessary time wondering whether or not a particular building should be included in the area or not, which is rather inefficient.

Scrawling (drawing freehand) an area proved to be intuitive and adapted (Figure 1d). Nonetheless, the problem is then translated at the implementation level, given the aim to produce a web responsive survey and that the desktop with a keyboard and a mouse has nothing in common with the gestures available on mobile.

Also, when respondents have the freedom to draw either points, circles, polygon or scrawled areas, it may be difficult for the analyst to infer trustworthy results from potentially so different answers.

Because of all the drawbacks listed above, the approach described below seems more appropriate and less risky.

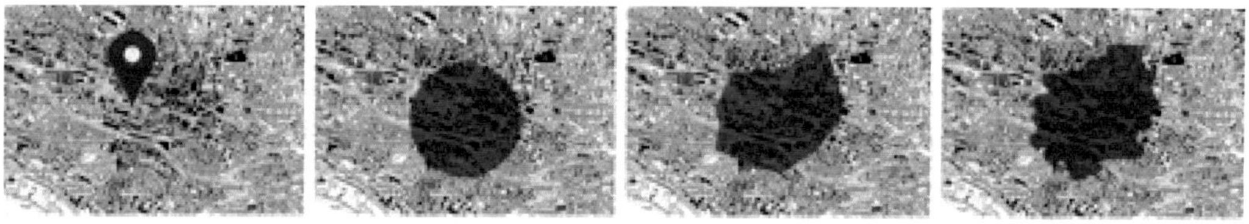

Figure 1: From left to right (a) Point (b) Circle (c) Polygon area (d) Scrawl (freehand).

3.2. Input by selection

With this second approach, the answer is a "portion of territory" which is a selection of one or many choices made by the user on the base of pre-existing entities. At a first glance, the expected interaction can indifferently respond to the logics of a single choice (radio- button) or a multiple choice (check-box). In the context of our case studies, the selectable entities are only polygons. Therefore, for questions of type "where do you live?", it will be necessary to define a layer of entities - in our case a regular grid covering the building areas - and to ensure that the scope of the selectable entities represents a good compromise between:

- User: a threshold in term of privacy perceived as acceptable by the user
- Analyst: a precision as sought by the survey designer
- Developer: a reasonable weight of the data to embed in the survey web page

4. From a Page Layout to a Web Framework

Let's consider the following question: we want to know in which area(s) the respondent does practice certain activities. With a survey in paper format and to make the best use of the layout, a list of all the possible answers may be presented just below the map (Figure 2).

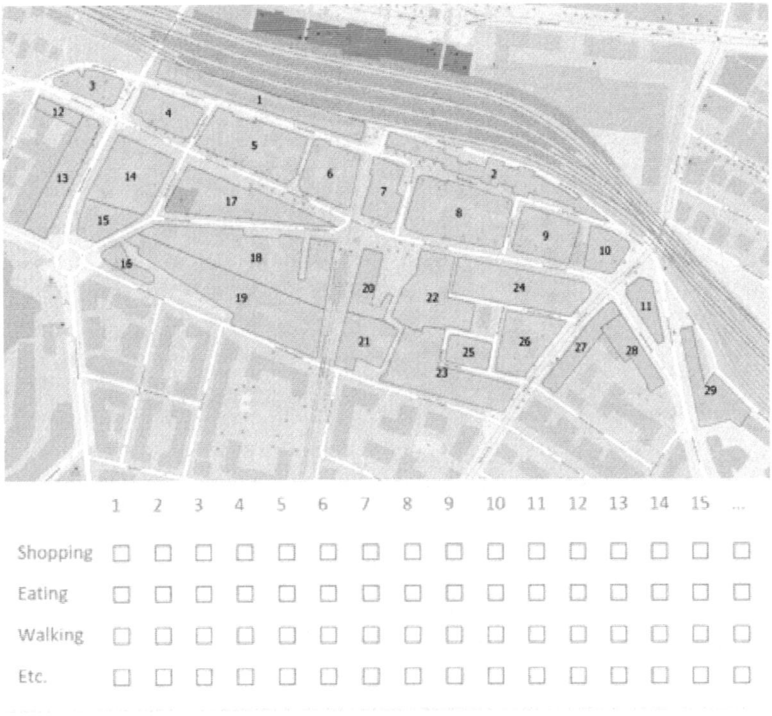

Figure 2: Example of a cartographic question for paper format.

This arrangement fits well on paper since the respondent has the question and the related map right in front of her/his eyes. But it is not advisable to transfer this logic as is on the web. Of course, it is possible to simultaneously display side-by-side a map and a list of check-boxes on a standard computer screen, but in terms of user experience it would be much nicer to interact directly with the map to respond. In this way, the numbering used for paper format would no longer have any reason to be. The mental effort could then be concentrated on the question itself rather than on how entities of the map are linked to the check-boxes.

Also, given how small and diverse the screens of smartphones or tablets are, definitely the paper model has to be avoided. We cannot reduce the size of the image without the risk of losing legibility or ask the user to scroll the page all the time to alternate between the map and the check-boxes. Therefore it is important to think differently.

Since we're talking about cartographic issues, a JavaScript cartographic library has to be chosen so as to go beyond the simple static image. The three most well-known libraries are probably Google Maps API, Leaflet and OpenLayers. As a research institute attached to a public university, we tend to favor open source technologies for the advantages they offer both in terms of education and applied research (Ertz et al. 2014). Between OpenLayers and Leaflet, we usually chose the first one (currently in version 3) because it is the framework we know best and it is also with this framework that we offer various courses to our students in geo-information (Geoinformation 2017).

With this library, we can easily produce a map of the size of our choice, centered on the position of our choice and with standard finger gestures already included (zoom, pan). This responsiveness solves the size issue of the map discussed above.

5. Geographical Data Best Practices

Any good webmapping library can add vector features on a map. These can be provided in multiple formats, GeoJSON, GML, GPX, etc. We chose the GeoJSON format because it is based on JSON which is lightweight and supported by many JavaScript libraries. Like jQuery, an useful library for addressing non-geographic aspects of a survey.

As soon as the entities are defined by the urban planner or extracted from a GIS (e.g. the regular grid covering the building areas), they have to be encoded with the GeoJSON format. We associate an identifier with each object, usually a unique attribute is enough to serve the three following needs: to meet the requirements of the analyst, to allow the identification on the map, to optimize the weight of the GeoJSON file.

OpenLayers allows to overlay features (e.g. polygons) transmitted through a GeoJSON file and to associate them with a style (Figure 3). The definition of the fill and stroke visual variables should avoid any confusion between the base layer of reference (which also requires the greatest attention) and the overlayed features.

Also some other aspects must be considered when creating GeoJSON files. In Figure 4 the cartographer proposes a relatively tight regular grid cut along the borders of the city of Lausanne. The aesthetic aspect is of a nice effect. But some areas on the edge are so small that they may contain only one building and then the user privacy problem mentioned above is encountered again. In addition, the small size of the areas on the edge forces the user to zoom in to ensure that she/he is ready to designate the right entity. The weight of the file is considerable (more than 500KB), which is not reasonable given the number of questions to provide or given the weight of other files (JavaScript, CSS, HTML and images).

By choosing a wider grid (Figure 5), we assume that the loss of precision results in a gain in efficiency, at least from the point of view of the user. Similarly, by presenting the entities in their raw form (not cut out according to the municipality perimeter), by reducing the number of decimal places on the coordinates of the GeoJSON file, and in some cases by using a topological format (e.g. TopoJSON), we come to an extremely compact file which speed up the loading time. An ultimate refinement would be to eliminate areas that are not relevant for the question (e.g. forest areas when requesting the place of residence) with a double benefit, decreasing the weight of the file and improving the legibility for the user.

Figure 3: Example of GeoJSON integration with OL3 (customized OSM base layer).

Figure 4: Without applying the guidelines (example of a regular grid covering the municipality of Lausanne).

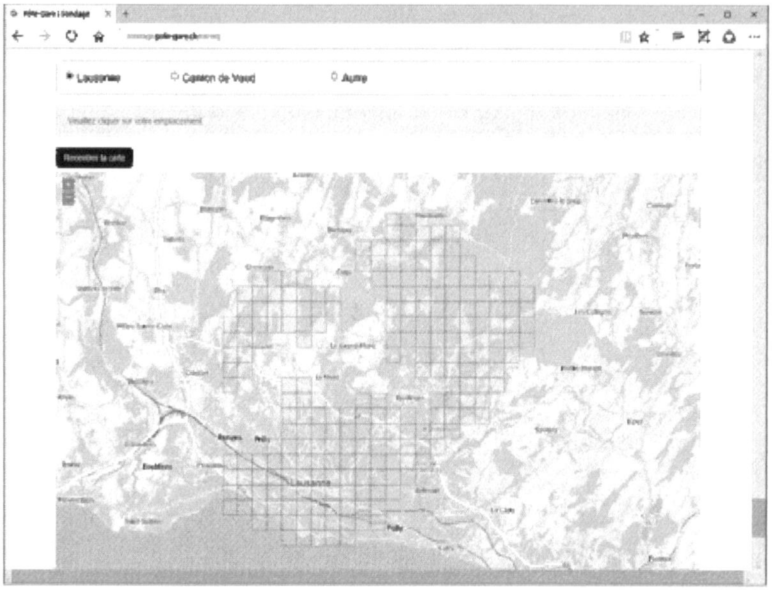

Figure 5: Considering the guidelines (example with all best practices applied to the city of Lausanne).

6. From Interaction Design to Implementation

In the previous section, the problem has been simplified to such an extent that it is now only about questions to answer by clicking on square areas on a map. Except the unknown ability of the average user to read a map, the proposed design is certainly efficient from the point of view of the user experience. There is no longer any intermediate step to link the check-boxes with the identified square areas on the map. In the following, details are given on the implementation and how to solve problems which ensue from.

At first, the user has to get a feedback as soon as she/he clicks a feature. This is essential to differentiate the selected features from those that are not. For instance, OpenLayers makes it easy to give a different style to features when they have been selected (Figure 6). Next point is then to find a way to gather the features selected by the respondent so as to be easily sent to the back-office. Our technical approach has been to address this problem in two parts: first, and given that each feature has been created on the map with a unique identifier, it is to retrieve the identifiers of the selected features so as to create HTML check-box elements, each holding one of the identifiers; secondly these elements are configured to be hidden in the user interface but ready to be sent to the back-office when the form is submitted. Regarding the back-office, this represents a quite loosely-coupled approach allowing to use any server-side framework. We have chosen to use Ruby on Rails to process the data and store them with PostgreSQL database system. The source code of both the front-office and the back-office is available online (MediaComem/survey-Pole-Gare 2017).

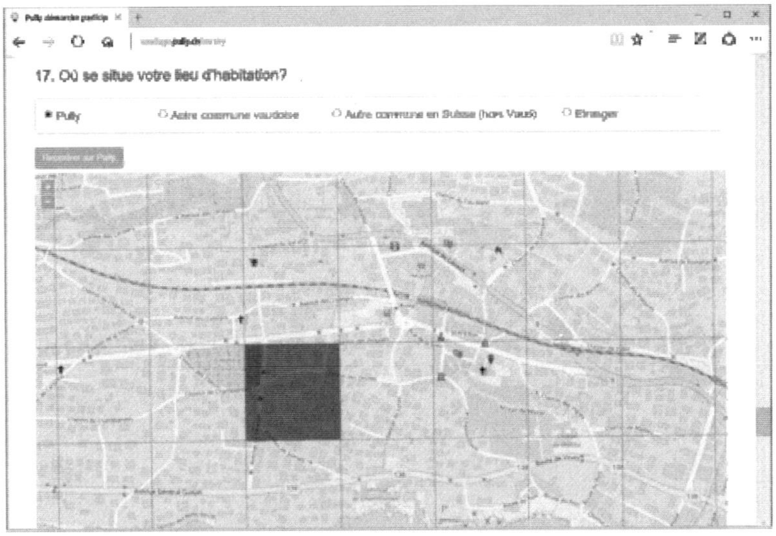

Figure 6: Example of single choice selection for the personal address (city of Pully).

Among the good practices discussed above to produce GeoJSON files, one is to find the right compromise between precision and efficiency, trying to control the number of features to be displayed. It reduces the loading time but also reduces the time needed by the user to answer. However, in the case of multiple choices, it is also necessary to avoid having to click on dozens of elements, which can be time consuming. But if this were to happen, we could further reduce the user effort by offering buttons to select or deselect all features with a single click. With the example presented in Figure 7, there are 20 selectable features from which 13 are selected (in red). Via the "select all" button the user can achieve this operation in 8 clicks instead of 13.

During pre-production phase, user testing has shown that it was quite common to see users quickly scroll the survey to skim through it. By default, on a computer, the more convenient way to zoom on a map and the action of scrolling a page are done in the same way, via the mouse wheel. And on a smartphone or tablet, the action of scrolling a page correspond to a pan on the map. As a consequence, the user may be easily disturbed when the intent is to scroll but has pan or zoom as a result. And what is even more disturbing is to see the zoom level or the initial position of the map involuntarily altered.

To solve this, we may think to disable the default behavior of the map during an action of scrolling the page. However, the problem persists if the action of scrolling does start from the place in the page where the map is presented. It is therefore necessary to calculate the dimensions of the map so that a margin remains on the screen, that is a space used as a starting point of the action of scrolling, especially on the tablet and smartphone. The Google Maps API offers an interesting solution by completely disabling the default action of pan on a map. The desktop user is then forced to use the zoom buttons and the mobile user to move on the map with two fingers instead of one. This solution seems to be the most appropriate, although you should at least be Google to impose a not necessarily natural and common practice to a user community!

Figure 7: Buttons to help the user for selecting and deselecting features.

Finally, other user testing does show how it is useful to have a map centering button that allows the user to reset the zoom level and the extent of the map when the context is lost when navigating. Also, in order to focus the users' attention, it is advisable to limit the overall extent of the map to the maximum (Figure 8). In this way, it is not possible to move the map too far outside the concerned geographical area (e.g. 10km around the bounding box of a city).

Figure 8: Maximum extent for Lausanne city. Only the red areas are useful for the question.

7. Practical Results

To complete the many guidelines for the design of usable web form (Bargas-Avila et al. 2010), we suggest the following recommendations when it is essential to incorporate a cartographic question in an online survey. These are validated empirically by means of the experience gained through the two case studies with the municipalities of Pully and Lausanne.

From general considerations to the specific design of the cartographic input type, the following guidelines are proposed:

Guideline 1	Consider the data analyst requirements in terms of data precision, usage and analysis
Guideline 2	Consider the respondent privacy and the time to respond
Guideline 3	Use a Javascript cartography library to display the map.
Guideline 4	Build a grid (vector layer) and use input by selection approach to collect answers from the map.
Guideline 5	Consider the stroke and filling colors of the grid to avoid any confusion between the map and the grid and/or adapt the base map layer.
Guideline 6	Find the good compromise between precision and efficiency to limit the number of the grid areas as much as possible.
Guideline 7	Do not create grid areas too small (or choose conscientiously the areas size to fit the respondent and data analyst needs).
Guideline 8	Limit the coordinates number of decimals and in some cases, it may be relevant to consider the TopoJSON encoding format to compact the vector file.
Guideline 9	Eliminate the areas that are not relevant for the question (e.g. pieces of lake, forest, etc).
Guideline 10	Obviously, use different colors to differentiate selected features from those that are not.
Guideline 11	Use traditional hidden form elements like checkboxes and radio buttons and bind them to the cartographic grid using unique identifiers.
Guideline 12	Show select all and unselect all areas buttons to help respondent.
Guideline 13	Show recenter button to help respondent to reset zoom level and extent of the map.
Guideline 14	Disable the default action of pan on a map and invite the desktop user to control navigation with the zoom buttons and the mobile user to move on the map with two fingers instead of one.

Moreover, the latest survey conducted with such cartographic questions has drawn up a list of 1530 responses, about half of which came from mobile devices. Only 63 responses (4%) remained unanswered for the cartographic questions. From which a little more than the half have been filled from smartphones or tablets. It is hard to say whether these questions remained unanswered is due to understanding problems (difficulties to read the map or to use the selection mechanism) or because the respondent simply did not want to answer. However, it is very likely that for a large part of these unanswered questions are related to the user's browser. Indeed, log analysis shows that 42 surveys in the 63 unanswered come from old browsers (Safari <6, Android <4 or IE <9) which can be problematic with the technologies used. It is therefore highly likely that the cartographic questions did not get displayed at all or at least not correctly in these cases.

8. Conclusion

The work carried out around these cartographic questions was successfully able to improve the user experience while preserving a traditional way of processing the results (simple scores are associated with the different geographical entities) for the analyst. It is also to notice that many respondents did communicate their significant satisfaction to be involved in such an participatory initiative. The good quality of the designed web form did probably participate to this positive judgment.

At the same time, some guidelines have been extracted as expected. This is to such an extent that all these good practices of design are now ready to be integrated into a specific library to be created and dedicated to cartographic questions for an interactive survey form. This would be a logical follow-up to the project. As a public funded research institute, the release of such a library would be done following the vision of Ertz et al. 2014. Indeed, they consider source code as text to share just as any research documentation, in other words, open source as a necessary condition to enable reproducibility, a core principle of science.

Also, beside the many guidelines for usable web form design that were applied in this work, it would be desirable to validate all these usability recommendations concerning the cartographic questions by pushing forward user testing, with eye-tracking in particular (M et al. 2014).

References

Bargas-Avila, J. A., Brenzikofer, O., Roth, S. P., Tuch, A. N., Orsini, S., Opwis, K., 2010. Simple but crucial user interfaces in the world wide web: Introducing 20 guidelines for usable web form design. User Interfaces.

Ertz, O., Rey, S., Joost, S., 2014. The open source dynamics in geospatial research and education. ournal of Spatial Information Science.

Geoinformation, 2017. Geoinformation course. Online, online; accessed May 31, 2017.
URL `http://mediamaps.ch/doku.php?id=geoinf`

Haklay, M., 2013. Citizen science and volunteered geographic information: Overview and typology of participation. Crowdsourcing Geographic Knowledge, 105122.

Ingensand, J., Nappez, M., Joost, S., Widmer, I., Ertz, O., Rappo, D., 2015. The urbangene project: Experience from a crowdsourced mapping campaign. 1st International Conference on Geographical Information Systems Theory, Applications and Management (GISTAM) 1-7.

Lausanne, 2017. Pole gare, la nouvelle gare et ses alentours. Online, online; accessed July 7, 2017.
URL `http://pole-gare.ch/`

M, S., Heinz, S., Bargas-Avila, J., Opwis, K., Tuch, A., 2014. Designing usable web forms: Empirical evaluation of web form improvement guidelines. Proceedings of the 32Nd Annual ACM Conference on Human Factors in Computing Systems, 12751284.

MediaComem/survey-Pole-Gare, 2017. Cartographic survey. Online, online; accessed May 31, 2017.
URL `https://github.com/MediaComem/survey-`

Pully, 2017. Le coeur de la ville au coeur du sujet. Online, online; accessed July 9, 2017.
URL `https://projets.pully.ch/fr/home/`

Modeling Effective Albedo as a Function of Land Cover Type and Snow Type

Martin Album Ytre-Eide[a,b,*], Terje Christensen[a,b], Tore Sivert Johnsen Istad[a], Lill Tove Nilsen[a], Bjørn Johnsen[a,b]

[a] *Norwegian Radiation Protection Authority*

[b] *Centre for Environmental Radioactivity (CERAD)*

Abstract: The Center for Environmental Radioactivity (CERAD) project UV- maps aims to obtain geographically distributed time-series of solar ultraviolet (UV) radiation in Norway. Since UV-measurements are limited to a few monitoring stations, a full representation of the spatial and temporal distribution needs to be based on a radiative transfer model (RTM). A key parameter is the regional albedo distribution. The albedo model developed here use a gridded set of local albedo values to derive the regional, effective albedo at any given point. In Norway there is a UV-monitoring network that has been operating since 1996, de- livering almost continuous 20-years time series of UV data, and the stations are used as reference points for the model. The albedo model uses land cover information and snow dispersion data from 11 years. Land cover classifications combined with snow classifica- tions constitute a matrix of albedo values, with one albedo value for each combination of land cover type and snow type, under the hypothesis that the snow albedo is affected by the underlying land type.

*Corresponding author
Email address: `martin.album.ytre-eide@nrpa.no` (Martin Album Ytre-Eide)

1. Introduction

Solar radiation is a primary factor for all life on earth. Organisms have developed strategies to take advantage of sunlight and at the same time manage harmful effects from UV-radiation. Disturbance in the solar radiation climate and climate induced changes in land cover, might lead to different species composition in the environment (Bjørn 2015).

Solar radiation has a temporal variation during the day and throughout the year and is influenced by many factors from topography and vegetation, to atmosphere and weather conditions. In many cases, like solar energy applications, climate, health and environmental studies, knowledge of geographically and yearly distribution of solar radiation will be relevant and important.

The reflecting power of a surface, albedo, is measured as the ratio of upwelling reflected radiation from the ground to the downwelling radiation hitting the ground, integrated for all angles of incidence (Feister and Grewe 1995). The albedo of a surface is generally spectrally dependent and in the following, only the UV part will be considered. Different surfaces have different albedo values and typical values are 0.02-0.05 for soil and vegetation (Feister and Grewe 1995), while for glaciated regions with homogeneous snow surfaces, such as Antarctica and Greenland, the surface UV albedo could be nearly 1.0 (Wuttke and Seckmeyer 2006). A part of the upwelling radiation from the ground is backscattered by molecules and particles in the atmosphere, resulting in an enhanced downwelling irradiance from the sky. We refer to this multiply scattered radiation as the diffuse contribution. If the albedo value is high, the diffuse contribution could be as high as 40 % for clear sky conditions (Wuttke and Seckmeyer 2006) , whereas in combination with clouds, enhancement values of up to 63 % have been observed of the total UV irradiation (Kylling et al. 2000). Knowledge of snow cover and snow quality is therefore of great importance.

Based on different land cover (sea, lakes, fjords, mountains, vegetation and urban environment), where we know the areal cover within a circle around each point, together with a time series of categorized snow dispersion, we hypothesize that the effective albedo, which is the sum off all influence from surrounding surfaces at a point, can be determined as a weighted function of area type, areal cover and distance to point. This hypotheses assumes that the aldedo value is known for all combinations of land cover and snow conditions.

2. Methods

2.1. UV network

In Norway there is a UV monitoring network of nine stations with 5-channel ground based UV radiometer (GUV) instruments (Johnsen et al. 2012). Since they have 2 channels in the UVB (280-315 nm) and 3 channels in the UVA (315-390 nm), it is possible to study different parts of the UV-spectrum. The stations have been working since 1996 — delivering a unique 20-year time-series of UV-data (Johnsen et al. 2012). They are spread across the country in order to give the best geographical and topographical representation of the UV radiation climate in Norway. One of the stations is located at Finse, a mountainous plateau in the south of Norway (60.6 degrees North and 7.5 degrees East). At about 1200 meter above sea level, Finse is snow covered from late October to mid May. The land covers here are a mix of open mountain, lakes and glaciers. When these areas are covered with snow, the area is quite homogeneous and is therefore suited for studies of how snow affects the effective albedo at the station. This study is therefore limited to Finse, but will later be extend to other stations.

2.2. Clear weather

Based on the observations at the UV-stations an algorithm for estimating the clearness index for all days was applied. The shape of the irradiation distribution over the day gave the UV-profile for that day, and python's scipy.optimize.curve_fit (Jones et al. 2001–) function was used to fit this curve. If the fit was good, the fitted values was used for comparison with the modeled values. Throughout this study only almost clear days were used. The choice of clear days is merely chosen as an instrument to estimate the albedo. Atmospheric conditions where the radiation can be scattered several times, in for example clouds, are not suitable for this type of calculation. Days with clouds are therefore discarded from the data.

2.3. From UV to albedo

The 380 nm GUV channel is sensitive to cloud interference, but not affected by ozone, and is therefor used to find cloud free days. If most of the day is clear, the clear-weather-UV-function fit is good. The Libradtran software (Mayer and Kylling 2005) performs radiative transfer calculations and is used to calculate a dummy 380 nm channel day-profile. In this calculation, the albedo was set to 0. The difference between the dummy profile and observations are due to the effective albedo, and the mode of the fraction and the relation between effective albedo and UV irradiation was used to obtain the observed effective albedo estimate.

2.4. Snow dispersion data

The Norwegian Water Resources and Energy Directorate (NVE) produces daily estimates of the snow situation in Norway, and has modeled snow conditions for at the last decade. NVE has a variety of snow dispersion data available and the chosen data that has four categories; "no snow", "some snow", "wet snow" and "dry snow", with spatial resolution of 1x1 km (Saloranta 2014).

2.5. Land cover

The N1000 area cover map from The Norwegian Mapping Authority (NMA) provides a categorical land cover vector map with key land cover surfaces. NMA share this map as a PostGIS dump file (Holl and Plum 2009), which also is our format of choice.

Table 1: The parameter matrix with albedo values to be determined

Land cover	No snow	Some snow	Wet snow	Dry snow
Glacier	a_{11}	a_{12}	a_{13}	a_{14}
Open	a_{21}	a_{22}	a_{23}	a_{24}
Water	a_{31}	a_{32}	a_{33}	a_{34}
Forest	a_{41}	a_{42}	a_{43}	a_{44}
Settlement	a_{51}	a_{52}	a_{53}	a_{54}

The format is downloaded as a zip-file and imported to local PostGIS server, and some of the categories are listed in table 1. This vector map covers all of Norway, but this study is focused on Finse, and therefore a 60 km x 60 km area around the Finse station is extracted by PostGIS's ST_Intersection and ST_MakeEnvelope functions.

A part from the great speed and possible command line interface, PostGIS access is implemented in a large number of GIS-softwares and programming languages. QGIS (QGIS Development Team 2009) is our chosen GIS-software to access our PosGIS database and convert the extracted vector map to a 100 m x 100 m raster map with the "Rasterize" function. The raster map is stored as a GeoTIFF file, because they are easily imported into python with the use of GDAL (GDAL Development Team 201x).

2.6. Albedo model

The albedo model aims to estimates the effective albedo at any given point in Norway. The model uses gridded, component surface albedo values. In table 1 the parameters that are relevant for the Finse area are listed. The parameter values are set for a raster map covering 60 km by 60 km area around the central point. Based on Walker 2009 the apriori assumption that the backscattered sky irradiance resulting from distant patches of snow, or snow free ground, is Gaussian declining from the point, a Gaussian distance weighting over that raster is applied. Summing the contributions from the surrounding areas gives the effective albedo in the central point. This method is illustrated in figure 1 and gives an effective albedo estimate which changes daily according to snow-conditions. Using the Finse station as central point means that a comparison of the model results with observations is possible.

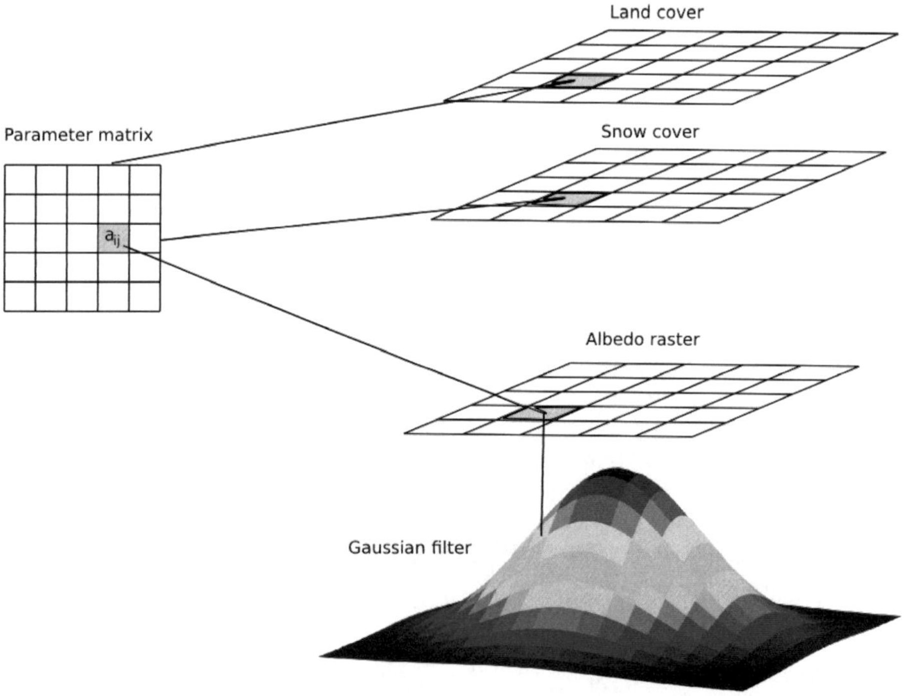

Figure 1: Illustration of the albedo model. Land cover and snow cover are used to find the aldedo value for each raster cell in an albedo map. A Gaussian filter is applied before summing the contributions to the effective albedo in the center.

2.7. Effective albedo equation

The effective albedo at point P can also be calculated as the sum

$$Albedo_{eff} = \sum_{i,j} a_{ij} x_{ij} \tag{1}$$

Here a_{ij} is the albedo parameter for land cover type i with snow type j, while x_{ij} is the effective area for land cover type i with snow type j. The effective area for each combination ij is calculated like this:

$$x_{ij} = \int s_{ij}(\vec{r}) g(\vec{r}) dA \tag{2}$$

where $\vec{r}$ is radius vector from P, and $s_{ij}(\vec{r})$ is 1 if position $\vec{r}$ is of type ij, and 0 otherwise, and $g(\vec{r})$ is a central symmetric Gaussian weighting function centered at P. The effective areas are all normalized so that their sum is 1.

3. Results

3.1. Parameter determination

The problem of determining the parameter values a_{ij} is solved by least squares regression using python's scipy.optimize least squares (Jones et al. 2001–) method. The bounds $min = 0$ and $max = 1$ was applied and the equation was solved with three different loss functions (linear, soft_l1 and cauchy).

The Finse area mainly consists of open mountain area, glaciers and lakes. At 20-30 km from the Finse station there are some forest areas, but the contributions to the effective albedo are assumed to be too small to be determined. Forests were therefore excluded and glacier's and lake's contributions where combined. Now there is four parameters to determine: Wet snow on open areas, dry snow on open areas, wet snow on lakes and glaciers, and dry snow on lakes and glaciers. The results are given in table 2.

Table 2: Albedo parameter regression results for Finse

Land cover	Wet snow	Dry snow
Glacier and lakes	0.93	1.0
Open	0.45	0.68

Table 3: Albedo parameter regression results for Finse, ignoring underlying land cover

Land cover	Wet snow	Dry snow
Finse area	0.52	0.72

As seen from table 2, the albedo parameter for dry snow with glacier or lake cover is pushed to the limit value 1. This is not a good estimate and the result is not physically rooted. The value for wet snow with glacier or lake cover is also higher than plausible. This discrepancy is probably because of the relatively small areas in combination with long distance to the center, resulting in a too low influence to be determined properly by the parameter regression method.

By ignoring the underlying land cover for Finse and only look at areas with different snow cover, wet or dry, the regression gives values shown in table 3.

The results seen in table 3 clearly distinguish between wet and dry snow and agrees with values found in literature (Feister and Grewe 1995). For glaciers and snow covered water, thees values may be too low and for open areas with rough stone formations these values seems too high. The vales found in table 2 is more in accordance to what would be expect for open mountain areas.

3.2. Model and observations comparison

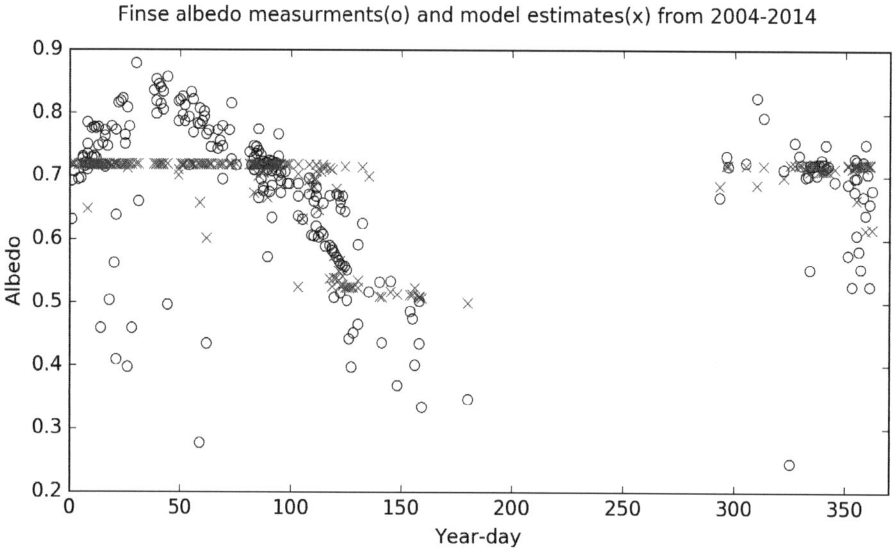

Figure 2: Clear weather effective albedo and model estimates at Finse UV-station.

The modeled effective albedo estimates, using the parameters found in table 3, are plotted in figure 2 where the circles are station observations and the crosses are model estimates. Only days when wet and dry snow covers more than 95% of the area are included in the figure. There is a systematic underestimation of the effective albedo from January to March, with a minimum at late January. From March to summer the differences are more spread, but with a mean closer to zero. The large differences seen between day 0 and day 50 are probably due to some random error or because of uncertainties in the effective albedo algorithm, which is highest during winter when the sun is low in the sky. The model captures some of the variation of the measured data, but there is clearly room for improvement.

4. Discussion

A description of a model for clear weather effective albedo calculations based on snow dispersion data and land cover has been presented together with an eleven year period of measured effective albedos at the Finse research station. Also, a methodology to determine the albedo parameters on which the model is based has been described. The current results from this parameter determination show that there are some issues that need to be addressed. Land covers which are not in the direct vicinity of the station tend to influence the effective albedo so little, that parameter determination is not possible and so these should be excluded from the analysis.

Figure 2 shows how the modeled effective albedos fits the observations at Finse UV-station. The modeled values follow the snow melting in April-May fairly well, although the modeled values seem a bit time lagged. In January and February, when the Finse area usually is covered with dry snow, the model is pushed to the upper albedo limit. Still, in late January/beginning of February, the model clearly undershoots the observation. This result shows that there are periods with albedo levels outside the bounds of the model. The reason for this might be because these periods have high abundance of snow and possibly also lower temperatures than at other times of the year. These are possible correlations which will be investigate in future studies. Deeper snow would make a better cover for the underlying rocky terrain and lower temperature would mean smaller snow grain size which have higher albedo. Also, this period is suspected to have more days with new snow, which would also increase the albedo levels.

The snow data used in this work have limitations, with only four categories and a resolution of 1 km x 1 km. More detailed snow data might make the albedo model more dynamic, but will also increase the number of parameters which need to be determined. The Gaussian distance weighting might also be a source of error, and it will in the extension of this work be developed a distance function based on the seasonal snow variation and effective albedo observation.

When albedo values are known, the influence of clouds on UV irradiance can also be included in the model. The modeled UV-irradianse can also be weighted by different types of action-spectra like the spectrum for erythema (de lEclairage 1999). It would also be possible to investigate the energy production of solar panels at different angels by applying spectral data of solar cells.

Although this project is ongoing and the results are preliminary, the methodology shows how FOSS4G technology can be applied in scientific research, to exploit the vast amount of spatial data which contributes to new and better scientific understanding.

Acknowledgments

The work was supported by the Research Council of Norway through its Centres of Excellence funding scheme project number 223268/F50.

We thank Jess Andersen at The Norwegian Water Resources and Energy Directorate for providing and assisting us with snow data.

We thank Tanya Helena Hevrøy for proofreading and correction of the language.

References

Bjørn, L. O., 2015. Photobiology - The science of light and life. Springer.

de lEclairage, C. I., 1999. Joint iso cie standard: Erythema reference action spectrum and standard erythema dose. iso 17166 cie s007 e-1999.

Feister, U., Grewe, R., 1995. Spectral albedo measurements in the uv and visible region over different types of surfaces. Photochem. Photobiol. 62 (4), 736–744.

GDAL Development Team, 201x. GDAL - Geospatial Data Abstraction Library, Version 2.1.0. Open Source Geospatial Foundation.
URL `http://www.gdal.org`

Holl, S., Plum, H., Apr. 2009. PostGIS. GeoInformatics 03/2009, 34–36.
URL `http://fluidbook.microdesign.nl/geoinformatics/03-2009/?page=34`

Johnsen, B., Nilsen, L. T., Dahlback, A., Edvardsen, K., Myhre, C. L., 2012. The norwegian uv-monitoring network: Qc and results for the period 1996-2011. AIP Conf. Proc. 1531 (1), 784–787.
URL `http://adsabs.harvard.edu/abs/2013AIPC.1531..784J`

Jones, E., Oliphant, T., Peterson, P., et al., 2001–. SciPy: Open source scientific tools for Python.
URL `http://www.scipy.org`

Kylling, A., Dahlback, A., Mayer, B., 2000. The effect of clouds and surface albedo on uv irradiances at a high latitude site. Geophysical Research Letters 27, 1411–1414.
URL http://dx.doi.org/10.1029/1999GL011015

Mayer, B., Kylling, A., 2005. Technical note: The libRadtran software package for radiative transfer calculations - description and examples of use. Atmospheric Chemistry and Physics 5, 1855–1877.

QGIS Development Team, 2009. QGIS Geographic Information System. Open Source Geospatial Foundation.
URL http://qgis.osgeo.org

Saloranta, T., 2014. New version (v.1.1.1) of the senorge snow model and snow maps for norway. Report no. 6.
URL http://publikasjoner.nve.no/rapport/2014/rapport2014_06.pdf

Walker, D., 2009. Cloud effects on erythemal uv radiation in a complex topography. Ph.D. thesis.

Wuttke, S., Seckmeyer, G., 2006. Spectral radiance and sky luminance in antarctica: a case study. Theor. Appl. Climatol. 85 (3), 131–148.

Open Geoportal lands to Europe: use cases and improvements from Geodata@Polimi

Marcella Samakovlija[a,*], Francesco Bartoli[b]

[a]*Leonardo Campus Library, Archive and Library Systems, Politecnico di Milano p.zza Leonardo Da Vinci, 32, Milano, Italy*

[b]*Geobeyond S.r.l. via Marchesa Augusta 68, 02040, Vacone (RI), Italy*

Abstract: TeDOC is part of the University Library System, owns a Great collection of historical and current maps, paper and digital and provides at their distribution to teachers and students, as part of the educational and research activities of the Politecnico di Milano. We were searching for a tool able to find, display and download current and historical cartography and, after a few years of work, we have become the first Europeans to use Open Geoportal (OGP) as our working tool. The component to ingest the data to OGP has evolved considerably, allowing better integration with GeoServer and allow automation of the procedures for adding layer via the importer function with the REST API. In this way, it is not necessary to be an expert operator to load data into the OGP. The procedure of data download has been customized to allow students to download prepackaged packages prepared by the administrators (some complex data need additional documentation to be correctly read, such as legends or lists of codes). One of our needs was to interact with other universities for sharing data, and OGP allows us to do this. For this reason, the Metadata that we have prepared follow the ISO 19139 standard and are in double language, allowing the widest possible dissemination and facilitating the exchange. What we hope is that our experience can become a first European base for the OGP Federation, also a new way to share data between Italian and European Universities.

*Corresponding author
Email address: marcella.samakovlija@polimi.it (Marcella Samakovlija)

1. Introduction

For a long time, we were looking for a tool that would be able to combine geographic research with the search for the most current cartographic typologies (raster, vector, scanned maps), but at the same time not exclude from research the historical cartographies that, for Italian copyright reasons, cannot be transformed into digital elements (Kollen et al. 2013). Usually, the historical cartographies information is contained in the classical University Catalogue (MARC21 and unimarc formats).

Open Geoportal (https://github.com/OpenGeoportal) is the tool we have chosen since it allows geographically searching of different typologies of maps in an easy and fast way. The community of Universities and Institutions linked to the Open Geoportal project has created this tool and all the necessary components in a balanced way, always following long debates, comparisons and experiments in a climate of mutual collaboration. We believe that sharing this knowledge from the community is crucial in identifying our closest issues and identifying the most appropriate solutions to overcome them.

Our needs in particular are:

- Allow an inexperienced operator to upload and process the data independently;
- Maintain the original format of the data as they were acquired by the owner Institutions;
- Allow our users to download the necessary data without necessarily going to the library;
- Allow to search and download data that our institution does not have directly and at the same time, allow sharing our cartographies with the federated Institutions (Muratori and Samakovlija 2015).

Figure 1: GeoData@Polimi

2. Advanced Integration with Geoserver

2.1. Exploiting the Importer plugin to automate the ingestion of raster and vector data

GeoServer is fundamentally one of the most important component in the architecture of Open Geoportal as it handles all the middleware OGC services for viewing and downloading functionalities. OGPIngest is the primary tool for end users that wants to upload data to Open Geoportal and the old procedure provides a two steps workflow to allow discovering vector and raster records from SOLR and make real data available from GeoServer:

- Load metadata from the OGPIngest GUI to the SOLR collection
- Create manually in GeoServer the associated resource, a layer, and the Web Mapping Service (WMS) or Web Feature Service (WFS) accordingly to what the previous step has declared in the location of its SOLR record.

The new OGPIngest is able to exploit a GeoServer equipped with the Importer plugin (http://docs.geoserver.org/latest/en/user/extensions/importer/) and use its REST API interface to automate the previous operations in a single step. Vector datasets such as shapefiles will be loaded as layer stored through a PostGIS datasource connection while raster ones such as JPEG and GeoTIFF will be treated accordingly to WorldImage and GeoTIFF datastore.

Our administrators, with proper privileges to load data in the geoportal, can indeed interact with the old option in the ingest GUI to select this automatic procedure as depicted in the Figure 2. At the end of each successful attempt, a new layer is immediately available and consumable within Open Geoportal for being searched, viewed and downloaded.

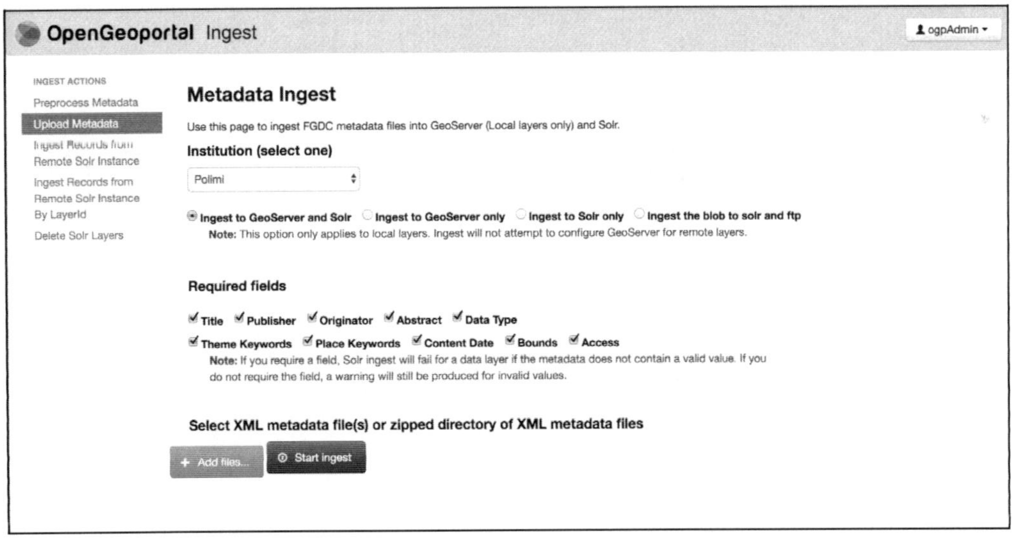

Figure 2: Upload menu of OGPIngest.

Behind the scene, OGPIngest performs a client HTTP call to the REST endpoint of the importer in order to accomplish the creation of the layer and the association with a specific datastore in GeoServer. These calls, for example in case of uploading a shapefile to PostGIS, are actually the same of the following commands:

```
curl -u admin:geoserver -XPOST -H "Content-type: application/json" -d @import.json
"http://localhost:8080/geoserver/rest/imports"
```

Figure 3: HTTP call example to trigger the importer.

```
{
  "import": {
    "targetStore": {
      "dataStore": {
        "name": "postgis"
      }
    },
    "targetWorkspace": {
      "workspace": {
        "name": "public"
      }
    }
  }
}
```

Figure 4: JSON file example to import a Shapefile, import.json.

The response will have an empty import task identification, which can be then used to trigger the real task:

```
curl -u admin:geoserver -F name=myshapefile -F filedata=@myshapefile.zip
"http://localhost:8080/geoserver/rest/imports/14/tasks"
```

Figure 5: HTTP call example to trigger a specific import task.

Then set the task to have a PostGIS target for the import action:

```
curl -u admin:geoserver -XPUT -H "Content-type: application/json" -d
@target.json "http://localhost:8080/geoserver/rest/imports/14/tasks/0/target"
```

Figure 6: HTTP call example to set the target for an import task.

```
{
  "dataStore": {
    "name":"postgis"
  }
}
```

Figure 7: JSON file example to set a target datastore, target.json.

Finally execute the complete action:

```
curl -u admin:geoserver -XPOST "http://localhost:8080/geoserver/rest/imports/14"
```

Figure 8: HTTP call example to execute the complete import job.

3. Management and Download of Blob Data

An improvement from the base Open Geoportal was planned to allow the end users to get assembled collection of generic files for the same datasets with few clicks. The new procedure has been required for handling any kind of blob data, usually a zip file with single or multiple vector/raster datasets correlated by any additional file for styles, data specification and standard metadata compliant to ISO 19139.

The solution has been designed toward a distributed architecture rather than a storage in a local file system. A new component - FTP/SFTP server - has been introduced to host the zip files and create the resource for the specific metadata field location in the Solr records. During the ingestion of blob the operator has a new option Ingest the blob to solr and ftp as depicted above in Figure 9.

The choice of FTP has been taken in order to meet the requirement of a reliable and robust way to exchange and host files with the most secure and consolidated protocol within a safe network zone. That doesn't mean it was selected also for the download procedure whose links will still continue to be served through the HTTP protocol and being protected by the identity and access management system. This also has the advantage to not require any additional integration of the FTP server within the perimeter of Single Sign-On performed by Shibboleth.

The final architecture can be described as per the diagram below:

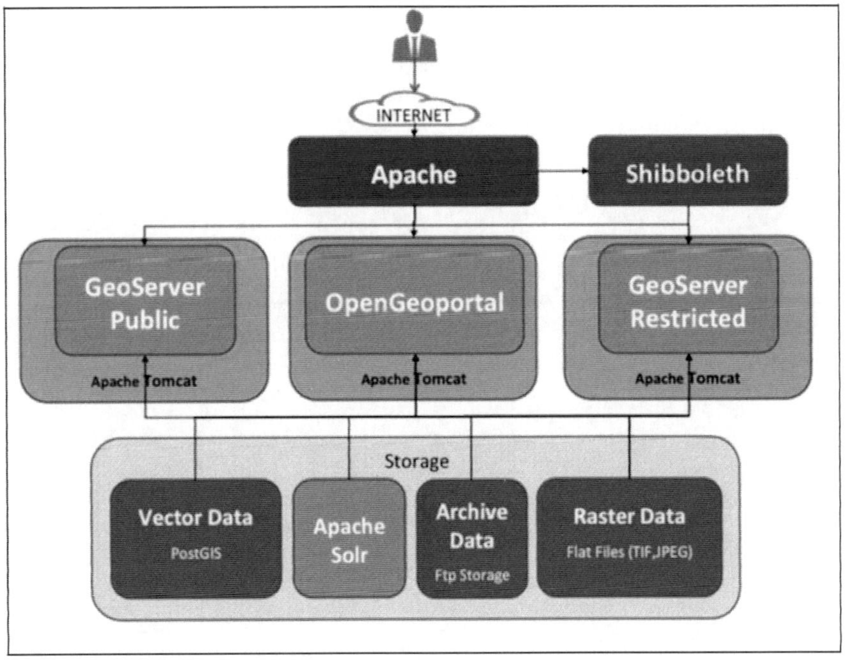

Figure 9: Geo Portal architecture for blob data.

3.1. Code improvement

To fill the solution described above an FTP client has been added in the OGPIngest and OGP controller to handling specific actions and trigger the movement of files from the upload to the storage and viceversa on users demand when they needs to download them asynchronously by email.

3.1.1. OGPIngest

A new bean for the FTP client is declared in the applicationContext.xml:

```
<beans:bean id="theFtpClient" class="org.OpenGeoPortal.Utilities.TheFtpClient">
        <beans:property name="ftpserver" value="${local.ftp.server}"/>
        <beans:property name="ftpport" value="${local.ftp.port}"/>
        <beans:property name="ftpusr" value="${local.ftp.usr}"/>
        <beans:property name="ftppwd" value="${local.ftp.pwd}"/>
        <beans:property name="ftpdir" value="${local.ftp.dir}"/>
</beans:bean>
```

Figure 10: OGPIngest application Context.xml with new bean.

And a corresponding properties section in the ingest.properties file:

```
local.ftp.server=192.168.56.102
local.ftp.port=21
local.ftp.usr=vagrant
local.ftp.pwd=vagrant
local.ftp.dir=/tmp
```

Figure 11: OGPIngest ingest.properties with ftp section.

3.1.2. Open Geoportal

A modified bean for the download method is declared in the *applicationContext.xml*:

```
<beans:bean id="downloadMethod.email.polimi"
    class="org.opengeoportal.download.methods.EmailDownloadMethodImpl"
    scope="prototype">
    <beans:property name="smtpuser"          value="${smtpuser}"  />
    <beans:property name="smtppassword"      value="${smtppassword}" />
    <beans:property name="smtpfrom"          value="${smtpfrom}"  />
    <beans:property name="smtpsubject"       value="${smtpsubject}" />
</beans:bean>
```

Figure 12: Open Geoportal applicationContext.xml with download bean.

And a corresponding json key/value file for configuring the download methods:

```
"polimi":[
      {"classKey": "layerDownloader.email.polimi",
"accessLevel":["public","restricted"],"dataType":["vector"],"outputFormats":["shp"]}
,
      {"classKey": "layerDownloader.email.polimi",
"accessLevel":["public","restricted"],"dataType":["raster"],"outputFormats":["geotiff"]}],
      {"classKey": "layerDownloader.email.polimi",
"accessLevel":["public","restricted"],"dataType":["vector","raster","undefined"],"outputFor
mats":["shp", "geotiff"]}],
```

Figure 13: Open Geoportal ogpDownloadConfig with configured classes.

4. Metadata in Double Language

The Politecnico di Milano has different maps with various typologies and provenances, the most recent ones are accompanied by a metadata according to the INSPIRE guidelines

(http://inspire.ec.europa.eu/documents/inspire-metadata-implementing-rules-technical-guidelines-based-en-iso-19115-and-en-iso-1) and therefore compatible with the ISO 19139 standard
(http://inspire.ec.europa.eu/id/document/tg/metadata-iso19139 and https://www.iso.org/standard/32557.html), while the maps of the 80s and 90s are free of metadata.

For all the native digital cartography, in order to allow the widest possible dissemination of the information contained (Goodchild 2000), the choice was to include abstract, lineage, keywords and usage in double language: Italian and English. The choice was also supported by the institutional bodies of the Politecnico di Milano who see in the internationalization one of the goals to be reached in the short term. In fact, more and more foreign students come to study in Italy and need to find suitable material for the tasks assigned.

For recent cartographies, in accordance with the guidelines drawn up by the federated (Florance et al. 2015) and in agreement with the owners of the data that could benefit from this work, we have prepared new metadata (always in double language).

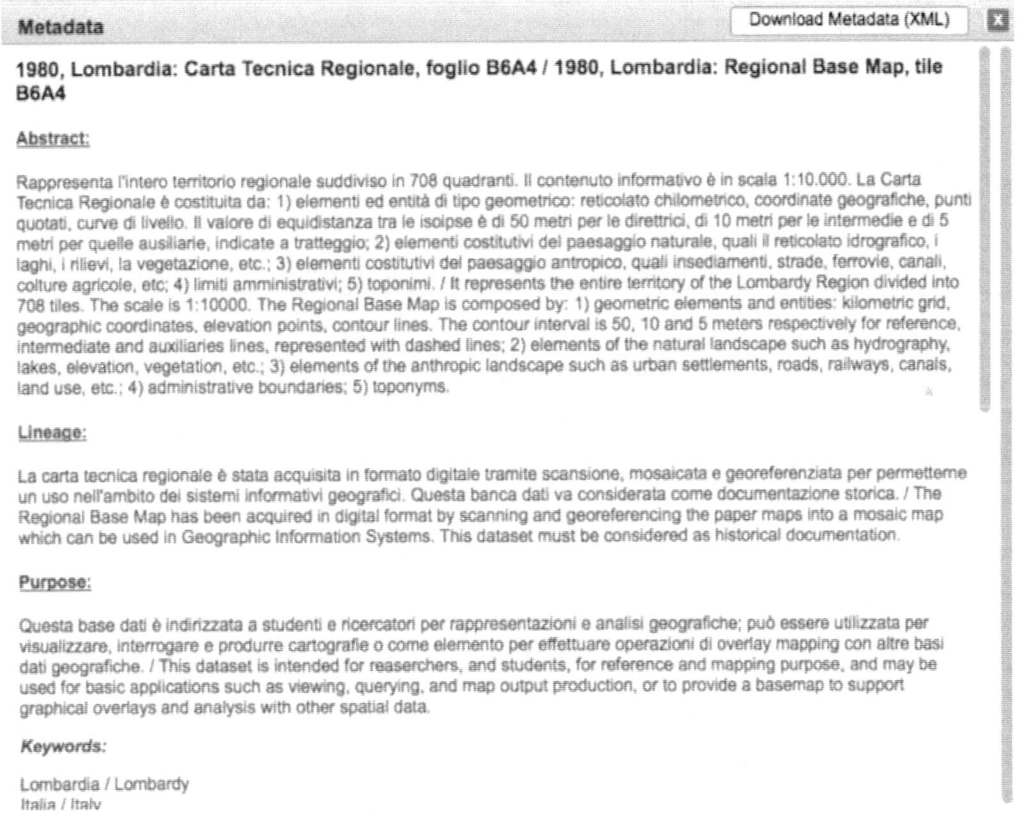

Figure 14: Example of metadata visualization in double language.

A series of cartographies, for copyright reasons, will never be available in digital format, but are already present in the University Catalogue. For this type of map, we followed the MIT guidelines to suit our needs. In this way the user can geographically search for this type of cartography even if cannot download them, but will have to go to the library personally. Technically, the catalog record is produced in Unimarc format (https://www.ifla.org/publications/unimarc-formats-and-related-documentation), converted to MARC21 (https://www.loc.gov/marc/bibliographic/) and extracted in xml format.

Through the tool MARCIngest (https://github.com/gravesm/marcingest) the MARC21 is inserted into the Solr instance allowing the user to easily search it. Within the catalogue card are inserted the geographic coordinates that allow to define the space occupied by the paper and to display its bounding box. The presence of the permalink allows the user to display the catalogue card directly inside the University Catalogue, where he will also find all the information necessary for his request.

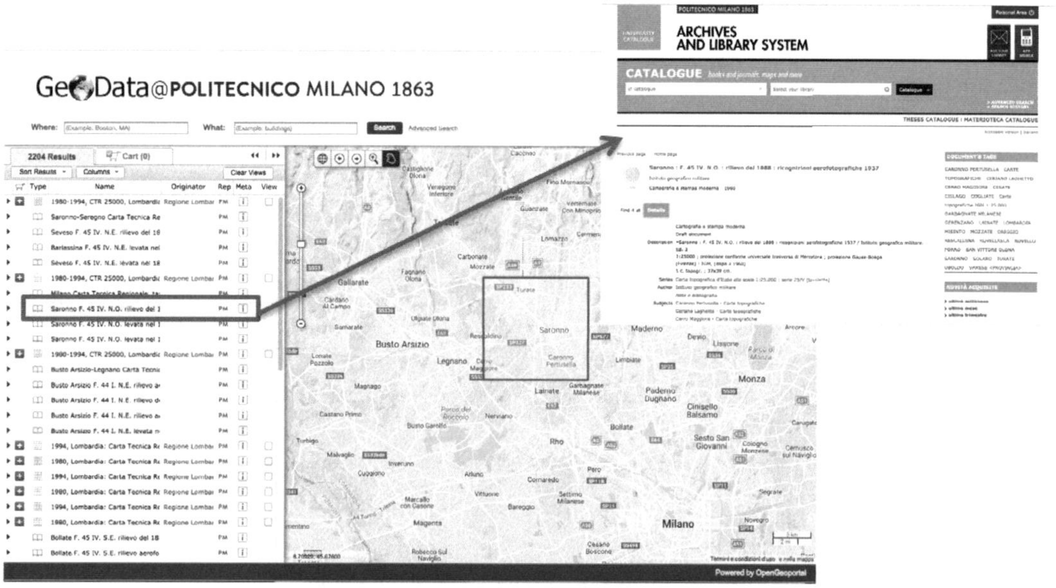

Figure 15: Example of paper map searching.

5. Conclusions and Next Development

The realization of GeoData@Polimi is not a point of arrival for us, but a starting point. We intend to make new improvements to OGP shared with the Federation, perhaps even having a multilingual interface (not just a metadata).

In particular, our work will focus on the opportunity of:

- digitalize and georeferencing historical maps owned by the Libraries (in respect of copyright);
- implement the data on the portal with the federation of other Institutions and Universities (Italian and foreign);
- use the geoportal to spread the georeferenced data produced by our Departments for study and research purposes.

The Politecnico di Milano has a rich collection of historical cartography, buried in the archives of Engineering, part of illustrious collections that for various reasons are difficult to find by users. It is our intention to start a work on this material, in order to make it available not only to the Politecnico users. Harvard colleagues, who have a very high-class historical map collection, will have the opportunity to suggest the most suitable method for scanning and georeferencing heterogeneous and ancient maps such ours.

GeoData@Polimi is not the first Open Geoportal in the world, but is the first in Europe and in Italy. We hope that our work can arouse the interest of other nearby universities so we can expand the Federation and share new materials. Finding cartography is getting easier through the geoportals of the various Institutions: each one with its authentication, downloading, or service modes (Florance 2006). For students, used to work on different geographic locations, having a single point for searching is an advantage. We would like to share this way of searching cartography with the students from other University as well.

Other partner in the Federation are working on how to build and evaluate the user experience of an Open Source geoportal5 (Wolff and Parker 2014), we are available to integrate the results of this research with our reference experience and the feed back provided by our users. We are at the beginning, but we count on the practice done by the Federates in these years of work.

Others Universities in Europe are partner in great projects such as Old Maps Online (http://www.oldmapsonline.org/) or promote directly their work on open linked data (http://lodum.de/). It could be very interesting find a way to explore these experiences and add new data and new interoperability option between this innovative projects and GeoData.

Like every University, our Departments and Laboratories produce quality research whose results are disseminated through international journals. It is difficult, however, to give the same relevance and dissemination to geographic data that has been realized from those researches (Durante and Wang 2012). Which best tool can be GeoData@Polimi? In addition to searching for basic geographic data, we will also make available those unique data produced by our researchers. We will have to work in collaborations with the Departments and the Federation to build appropriate metadata, which unequivocally provide the value of innovation.

References

Durante, K., Wang, Z., 2012. Creating an actionable assessment framework for discovery services in academic libraries. College Undergraduate Libraries 19(2-4), 215–228.

Florance, P., 2006. Gis collection development within an academic library. Library trends 55(2), 222–235.

Florance, P., McGee, M., Barnett, C., McDonald, S., 2015. The open geoportal federation. Journal of Map Geography Libraries 11, 376–394.

Goodchild, M. F., 2000. Communicating geographic information in a digital age. Annals of the Association of American Geographers 90(2), 344–355.

Kollen, C., Dietz, C., Suh, J., Lee, A., 2013. Geospatial data catalogs: Approaches by academic libraries. Journal of Map Geography Libraries 9, 276–295.

Muratori, I., Samakovlija, M., 2015. Sharing geodata through university libraries: the case of politecnico di milano. Online, online; accessed May 31, 2017.
URL `http://eprints.rclis.org/27995/`

Wolff, R., Parker, K., 2014. Usability testing: Open geoportal 2.0. place project. paper 9. Online, online; accessed May 31, 2017.
URL `http://scholars.unh.edu/place/9`

Optimizing Spatiotemporal Analysis using Multidimensional Indexing with GeoWave

Rich Fecher[a,*], Michael A. Whitby[a]

[a]*Digital Globe*

Abstract: The open source software GeoWave bridges the gap between geographic information systems and distributed computing. This is done by preserving locality of multidimensional data when indexing it into a single-dimensional key-value store, using Space Filling Curves (SFCs). This means that like values in each dimension are stored physically close together in the datastore. We demonstrate the efficiencies and benefits of the GeoWave indexing algorithm to store and query billions of spatiotemporal data points. We show how applying its multidimensional indexing strategies can reduce query and processing times by multiple orders of magnitude using publicly available taxi trip data published by the New York City Taxi Limousine Commission (NYCTLC). Furthermore, we demonstrate how this efficiency lends itself to analysis that would otherwise be unfeasible.

*Corresponding author
Email address: `richard.fecher@digitalglobe.com` (Rich Fecher)

1. Introduction

With the constantly growing amount of data available for use in geospatial applications, the ability to scale those applications is becoming more and more important. Distributed computing systems provide a highly scalable solution, but they also store data in single dimensional key-value stores that are not natively able to preserve multidimensional locality. To preserve multidimensional locality in a distributed key-value store is to store similar values in each dimension using similar keys, resulting in values physically close together to the great benefit of range-based retrieval. The open-source software GeoWave (GeoWave 2017) uses SFCs to efficiently index data into a key-value store in a way that preserves locality. The configurable options and expandable capabilities of GeoWave allow a user to adapt it to best fit his or her specific use case. In section three, we will show the benefits of GeoWave's indexing approach in the context of a real-world multidimensional use case. We measure a reduction in query and analytic times by multiple orders of magnitude over a technique that does not use GeoWave's locality preservation. In section four, we will discuss an application that utilizes billions of historical taxi trips indexed into the GeoWave framework to provide interactive trip estimation to a user.

2. Design Description

One of the greatest advantages of GeoWave over other current spatial and temporal indexing methods is the ability to work with any number of dimensions. The primary method to accomplish this is a tiered, hierarchical, SFC storage system. GeoWave uses Compact Hilbert SFCs (Hamilton and Rau-Chaplin 2008) to represent n-dimensional grids at multiple tiers that it then uses to provide an index in which each elbow (discrete point) of an SFC maps to a cell of the grid (Whitby et al. 2017). To achieve varying levels of precision, each dimension is progressively decomposed in a manner similar to a quadtree. However, there is a significant difference between the GeoWave process and a standard quadtree as there is no direct link between progressive levels (or tiers) in the GeoWave decomposition. Each entry in the datastore has its corresponding tier stored in its key as part of its Row ID (Whitby et al. 2017).

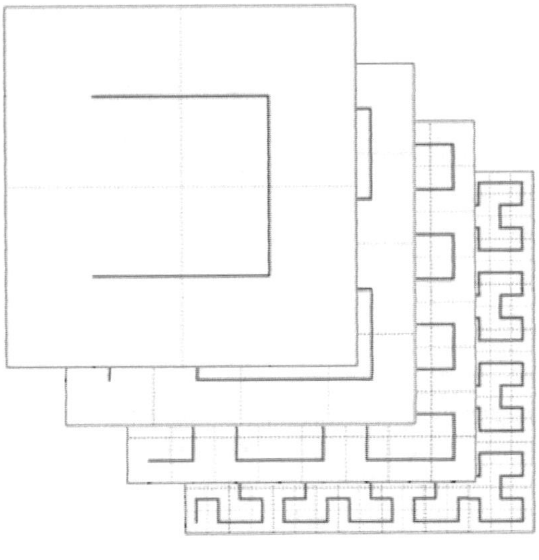

Figure 1: Visualization of the GeoWave tiered SFC storage system.

GeoWave also provides the ability to index multidimensional data into a key-value store using the XZ-Ordering algorithm proposed by Bohm (Bohm et al. 1999). The XZ-Ordering algorithm has the benefit of being insensitive to grid resolution. This ability, combined with the avoidance of object decomposition, makes it beneficial for use in indexing regular shaped line/polygon data.

A further challenge is that of unbounded dimensions like time. Unlike latitude and longitude, time does not have a standardized beginning or end, which would normally make it impossible to normalize in a way that would fit into an SFC. GeoWave solves this issue by defining a configurable periodicity for that dimension and creating bins for each period. Each bin is a hyperrectangle representing ranges of data and labeled by points on a Hilbert SFC (Whitby et al. 2017). A three-dimensional representation of this process is show in Figure 2.

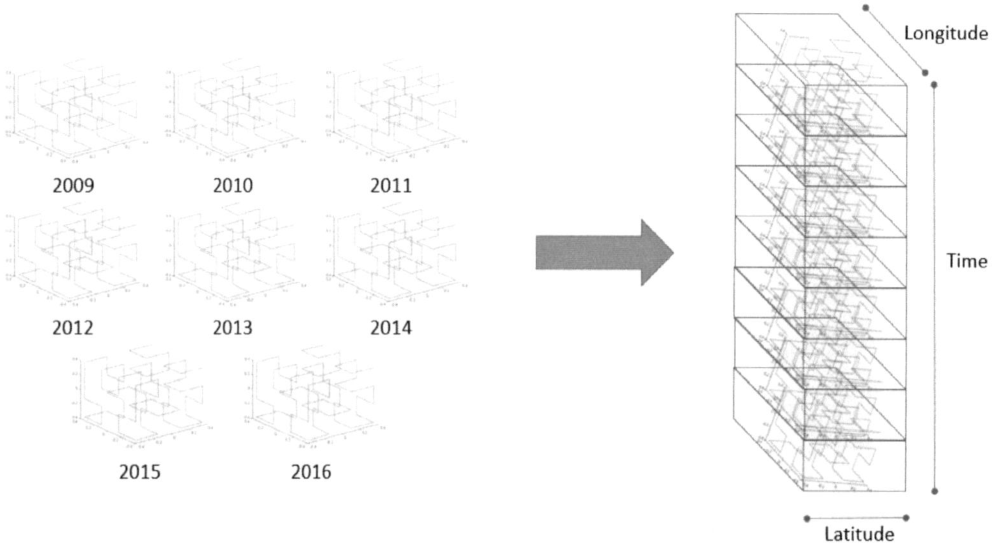

Figure 2: Composing the unbounded temporal dimension.

The GeoWave framework is implemented on top of a distributed key-value store like Apache HBase (Apache 2017b) or Apache Accumulo (Apache 2017a), however its user facing tools and developer facing APIs are agnostic to the underlying datastore (Whitby et al. 2017). This allows a user to take advantage of GeoWave's indexing strategies and analytics on whichever key-value store he or she deems best suited for the job.

3. Evaluation

For the experiments done in this section, we are using data provided by the NYCTLC on over 1.3 billion taxi trips dating back to 2009 (TLC 2016). We used GeoWave to index this data into an Apache HBase (Apache 2017b) backend in multiple ways to demonstrate the advantages and disadvantages of the various methods, as well as the level of customization that GeoWave provides. We also measured the performance of using HBase in this scenario without leveraging GeoWave's indexing approach. Table 1 provides the names of the indexing methods that will be referred to in this paper and shows which particular attributes, or dimensions, are indexed by each. The names of each method are intended to be partially descriptive in order to help differentiate between the methods in the results figures. The None method represents a full table scan using a unique Feature ID we created for this dataset. This None method represents a naive way of storing multi-dimensional data within HBase without any efficiency gains provided by GeoWave. The table also designates whether the method used the traditional tiered technique or the XZ-Order technique. Time Range in this case is defined as the time range between the pickup time and the drop-off time. Table 1 shows indexing methods used for the NYCTLC data.

Name	Pickup Latitude	Pickup Longitude	Pickup Time	Drop-off Latitude	Drop-off Longitude	Drop-off Time	Time Range
6D	X	X	X	X	X	X	
5Dtiered	X	X		X	X		X
5Dxz	X	X		X	X		X
3Ddropoff				X	X	X	
3Dpickup	X	X	X				
2Ddropoff				X	X		
2Dpickup	X	X					
None							

Table 1: Indexing methods used for the NYCTLC data.

In this set of tests, we want to know how many taxi trips were taken to and from Yankee Stadium during the 2009 baseball season, the playoffs, the World Series (plus the final game), and the subsequent off-season. The speed at which these queries can be performed when dealing with a dataset of over 1.3 billion features is greatly dependent on how the data is indexed.

Running these queries on this amount of data would normally take a very large cluster. The raw data alone is over two terabytes in size due to our requirement for the experiment to index it in so many ways. We were able to utilize a feature of the Apache HBase (Apache 2017b) integration with Amazon Web Services' (AWS) Elastic Map Reduce (EMR) to store our GeoWave tables in the AWS S3 object storage system. Since we decoupled our compute requirements from our storage requirement, we were able to run all of the queries on an EMR cluster with one m3.xlarge master node and three m3.xlarge worker nodes. Our compute requirements are relatively light for the better index methods we run range scans with efficient, well-defined ranges from GeoWave. But to theoretically allocate enough HDFS disk space for all of these tables, we would have required more than an order of magnitude increase in cluster size.

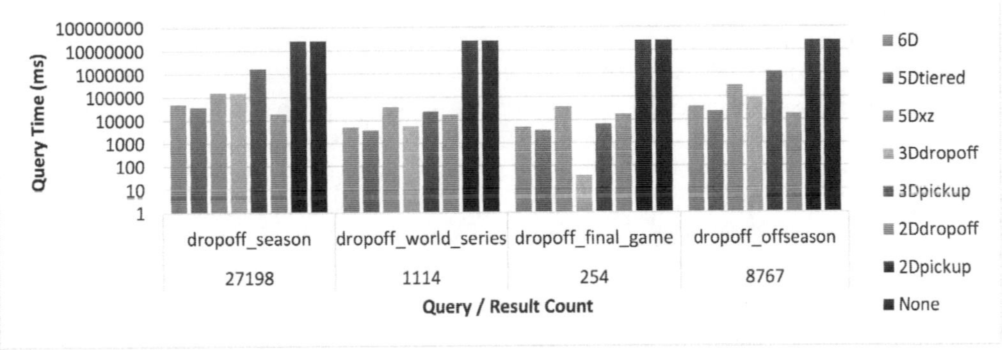

Figure 3: Results for taxi drop-off queries.

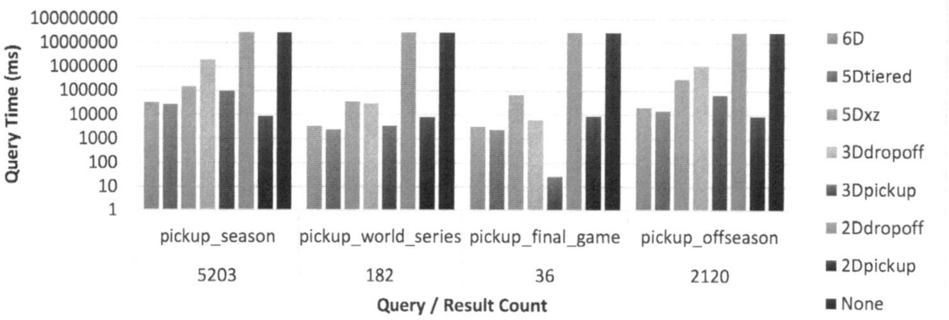

Figure 4: Results for taxi pickup queries.

As shown in Figures 3 and 4, a spatially and temporally constrained query has the best performance when the indexed attributes match the same latitude, longitude, and time as the query. This is why the 3Ddropoff index performs very well in the first query and why the 3Dpickup performs so well in the second query. However, the issue with both of these indexes is that their performance suffers dramatically when they are used for a query in which the spatial component of the index is not tightly constrained. In the case of the 3Ddropoff, the second query tightly constrains the pickup location but the drop-off location is completely unconstrained. This issue is even further exaggerated in the 2D indexes. The 3D indexes were indexed with respect to time, so even in the unconstrained spatial queries it still wasn't necessary to search though the entire data set. When using the 2Ddropoff indexed dataset for a query when the drop-off latitude and longitude are unconstrained, it forces the system to search though the entire dataset. This takes over seven and a half hours and defeats the purpose of indexing the data at all. This discrepancy between query times is visualized in Figure 6.

With a sufficiently sized dataset, it is neither efficient nor desirable to re-index the data for each individual query. The solution to this is to index the dataset with respect to a sufficient number of dimensions as to allow all of the expected query patterns to complete in a reasonable amount of time. We have done that with the 6D, 5Dtiered, and 5Dxz indexes. These indexes have similar performance between each of the eight queries (see Figure 5) with the 5Dtiered index performing best overall. The 5Dxz using the XZ-Ordering for its indexing method showed poorer performance as this is not the type of data for which it is optimal. We have found it performs best with more regular shaped extents that is data that has more similar lengths per dimension. In this case, the tiered indexing is the preferred method for point data such as this because we are able to decompose these queries to a very low level in the tiered structure, and only one of the dimensions, the time range, must be indexed as an extent. The 6D index was not optimal in this use case because our definition of the Time Range used in the 5Dtiered index was sufficient for the queries we ran here. If, within our queries, we needed to differentiate between taxi trips that were picked up within a certain time range and then dropped off within a separate time range the 6D index would be required.

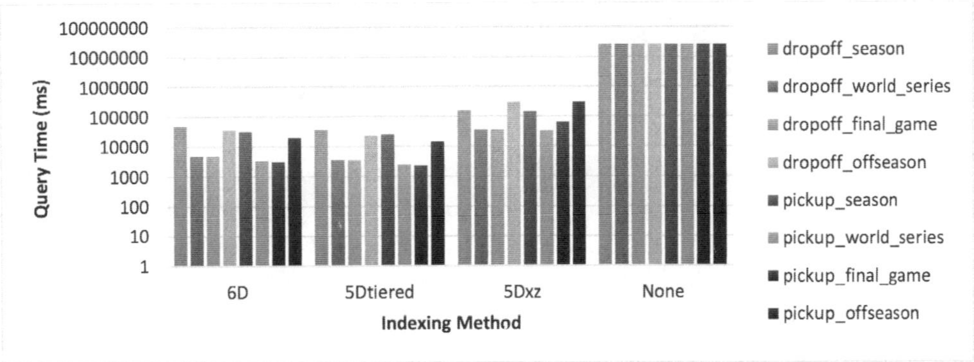

Figure 5: Results by indexing method.

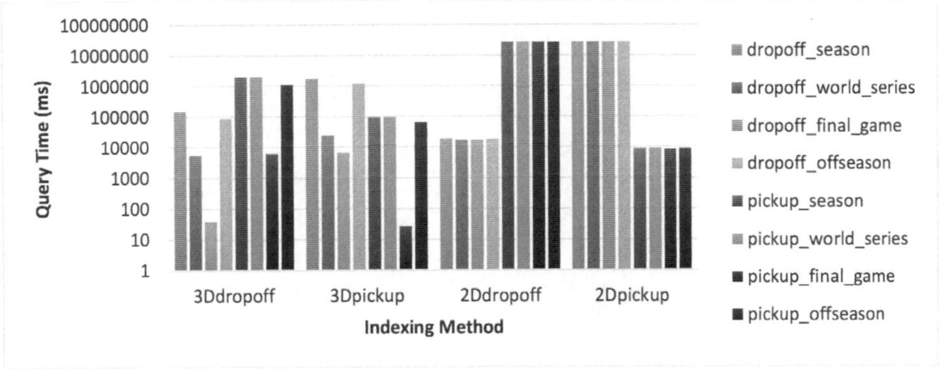

Figure 6: Results by indexing method (cont.).

In these basic data retrieval exercises, we show how a variety of questions can be asked of massive datasets at interactive rates given the appropriate indexing. Now we will extend this exercise of merely counting records to perform more advanced data analysis. We utilize the nature of distributed storage techniques lending themselves to large scale processing to run a parallelized, distributed Kernel Density Estimation (KDE) algorithm and ultimately produce heatmaps. We aggregate the data during each baseball season between 2009 and 2016 for any taxi trips that either started or finished at Yankee Stadium within this KDE process, producing weights for areas where passengers are either going to or coming from Yankee Stadium. We then run the KDE for the likewise trips during the offseason between the same years. These weights are scaled by a ratio of the number of total days during a baseball season to the total days of offseason and subtracted from the first set of weights. This step, including the offseason, normalized by the ratio of time periods, is intended to offset the theoretical regular taxi traffic during the season to and from Yankee Stadium that is completely unrelated to baseball games. This is based on an assumption that the traffic unrelated to games is fairly constant year-round. It is worth noting that most types of queries measured above are required to perform this analysis. Moreover, there are 34 distinct queries to account for both pick-ups and drop-offs at Yankee Stadium for each individual season and offseason time period. Therefore the 5Dtiered index was chosen as the data source to minimize worst case performance. We certainly wanted to avoid waiting several hours for any one query, and this index had proven flexible enough to retrieve the data for all of these queries within 30 seconds or much less. The final aggregate heatmap was added as an interactive layer on a webmap with screenshots shown in Figure 7.

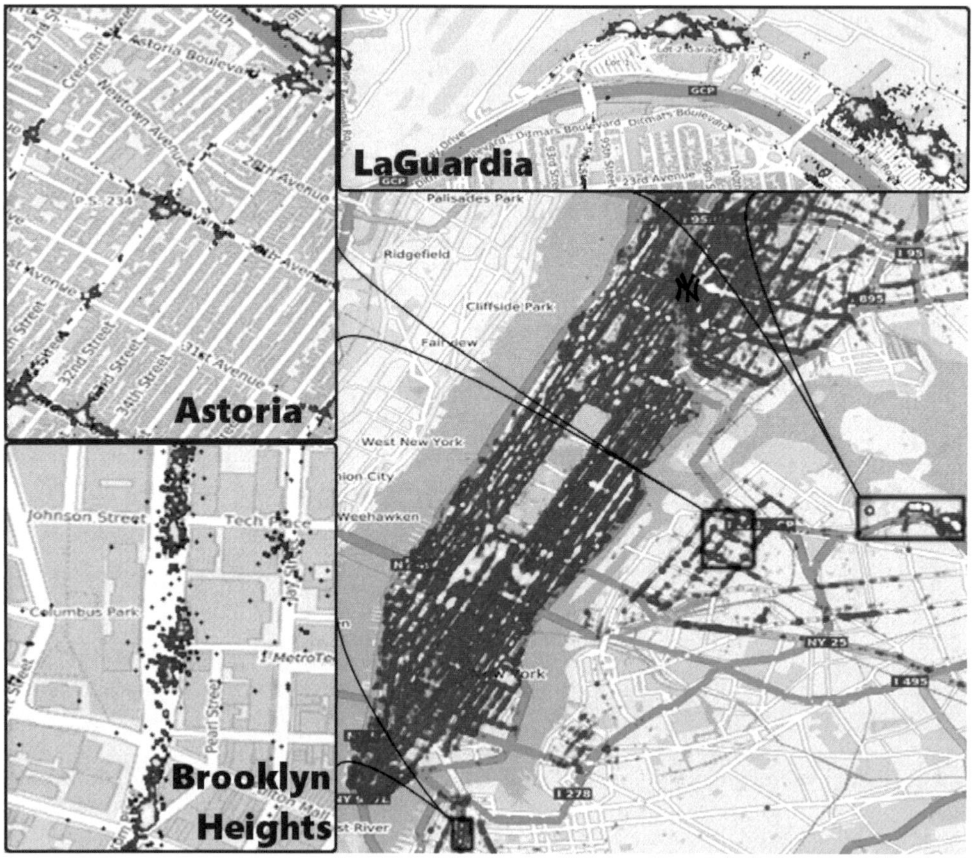

Figure 7: Overview heatmap and detailed area callouts Graphic background MapBox and OpenStreetMap.

Many of the trends seem intuitive, such as high amounts of taxi trips between Yankee Stadium and Manhattan, but some trends were not so apparent without taking a closer look.

There are hotspots of many taxi trips within Brooklyn Heights and Astoria, particularly near subway entrances. Without knowing New York City very well, this didn't make particular sense until we took a close look at the subway network itself. It is not possible to access the Yankee Stadium subway stop from these locations, particularly Astoria, without incurring transfers and a lengthy trip. The taxi ride is not far though, particularly between Astoria and Yankee Stadium. Clearly, baseball fans making a trip from the extremities of the subway system often find it easiest to get off of the subway without going into Manhattan, followed by leveraging taxi service for the final leg of the trip. Another result that initially caused confusion was significant activity at LaGuardia Airport. We did not realize baseball fans would go directly from the airport to the ballpark and back, but it is actually common. Many baseball fans fly in for a game, and use taxi services directly to and from LaGuardia Airport.

4. Demonstration

We extended the GeoWave framework into a project called nyc-taxi with an interface that allows a user to select a start and end point on a map of New York City for a theoretical taxi trip. We then created a 20-node EMR cluster. The nyc-taxi project was installed onto the cluster, and a GeoWave datastore with an Apache Accumulo (Apache 2017a) backend was created. The publicly available dataset with over 1.3 billion historical data points was ingested into the GeoWave datastore with locality preservation in the five following dimensions - time of day, pick-up latitude, pick-up longitude, drop-off latitude, and drop-off longitude (TLC 2016).

Figure 8: Screenshot of Taxi estimation demo.

The GeoWave indexing strategy allows the application to efficiently query the massive historical dataset for similar taxi trips based on user-selected start and end points, as well as the current time of day. The returned data can then be used to calculate useful information and display it for the user. The aggregation of these statistics is distributed within each data node so that it appears real-time to the user. Once the markers are placed on the interface map, the application displays the approximate trip length, average fair, estimated distance, etc. to the user interactively. As shown in this paper's experimental results, without the multidimensional locality preservation provided by the GeoWave framework, this application would have to perform much larger scans on each request, causing a significant reduction in its performance.

5. Conclusion

Using the GeoWave framework, we were able to show the massive benefits of using multidimensional indexing to preserve locality in a single dimensional key-value store. Specifically, we showed the benefits of n-dimensional indexing in regards to spatiotemporal applications and discussed the methods that GeoWave uses to accomplish it. We also demonstrated an application that extends the GeoWave framework to interactively return an estimated trip length and cost to a user based on over 1.3 billion historical taxi trips. GeoWave is an open source project and can be found on GitHub at https://github.com/locationtech/geowave.

References

Apache, 2017a. Apache accumulo. Online, online; accessed July 21, 2017.
URL `https://accumulo.apache.org`

Apache, 2017b. Apache hbase. Online, online; accessed July 21, 2017.
URL `https://hbase.apache.org`

Bohm, C., Klump, G., Kriegel, H.-P., 1999. Xz-ordering: A space-filling curve for objects with spatial extension. Proc. 6th SSD, LNCS 75-90.

GeoWave, 2017. Geowave. Online, online; accessed July 21, 2017.
URL `https://github.com/locationtech/geowave`

Hamilton, C., Rau-Chaplin, A., 2008. Compact hilbert indices: Space-filling curves for domains with unequal side lengths. Information Processing Letters 155163.

TLC, 2016. Tlc trip record data. nyc taxi limousine commission. Online, online; accessed July 21, 2017.
URL `http://www.nyc.gov/html/tlc/html/about/trip_record_data.shtml`

Whitby, M. A., Fecher, R., Bennight, C., 2017. Geowave: Utilizing distributed key-value stores for multidimensional data. 15th International symposium on Spatial Temporal Databases.

QKan - Management of drainage system data with QGIS

Jörg Höttges[a,*]

[a] *Lehrgebiet für Wasserwirtschaft und Bauinformatik, FH Aachen University of Applied Sciences 52066 Aachen, Bayernallee 9, Deutschland*

Abstract: Based on QGIS a database structure and a set of plugins have been developed to improve the workflow for the hydraulic design of urban drainage systems for consulting engineers. The main goal is pre- and postprocessing of the drainage system data in combination with various commercial hydraulic simulation software packages. The plugins provide import and export functionalities, simulation result viewing and generation of longitudinal cross sections in a CAD program. All modules make intensive use of SQL based spatial functions instead of functions implemented in QGIS, because of their stability, higher flexibility and speed due to the indexing functionality. On the other hand the database structure is kept simple to make it possible for engineers as a typical user to work directly with the data in the database.

*Corresponding author
Email address: `hoettges@fh-aachen.de` (Jörg Höttges)

1. Introduction

Since the release of version 2 of QGIS consulting engineers, water boards and public authorities in the Germany and Switzerland started to use QGIS due to its usability and the absence of license restrictions. With the upcoming of open data, which was slightly delayed in Germany for legal reasons, application of GIS is accelerated significantly. QGIS has become very useful for application development, because it offers all functionality that is necessary for consulting engineers working in the wide area of water management.

Whereas drainage system data is hold in database-driven network information systems operated by public authorities (cities, water boards) being the legal owner of the sewage systems, a lot of work has to be done to prepare the data for hydraulic simulation of stormwater runoff. For this purpose temporal modifications of the data are necessary, which unfortunately are not supported by the software systems used by the authorities. Hence the data has to be exported and to be modified in the simulation software package afterwards. For many years the developers of the simulation software packages have created their own user interface for this task, whereas during the last years they shifted to a GIS based user interface. One of the main reasons was the necessity of spatial analysis functions, e.g. for the collection of the fraction of pervious and impervious sub-areas.

In contrast to the commercial GIS based applications and also to other QGIS based application modules like the swiss QGEP project (Fischer 2015), the author of this paper gathered the experience that the best way of working for consulting engineers is to have an open database structure which can be modified to satisfy individual needs in contrast to a closed system with fixed and predefined workflows. Consultant engineers in Germany usually work for various authorities, which means they have to be more flexible and adapt to various data structures and sometimes also use different simulation software packages. On the other hand, it can be very useful to establish an engineering team workflow partially independent from the chosen simulation software. This has been an important design criteria for the QKan plugins.

2. Database Structure

QKan uses a SpatiaLite database. As an alternative a POSTGIS database will be implemented with an option for the user to choose between the two. It is very efficient to handle the table data via SQL statements (Figure 1) using the QGIS DB Manager or the spatialite-gui tool (Furieri 2017). Therefore the database omits the declaration of referential integrity using additional primary and foreign keys, but uses direct references between label attributes in the base table and the parent table instead. The disadvantage of lower performance and the abandonment of a verification of uniqueness is compensated by the fact that SQL statements prove to be much simpler and thus enable an easier approach to the user. Usually the uniqueness of labels is reviewed by the engineer during the export process to the simulation software.

3. Geo Functions Database-Driven

The author has come to the decision that using the SQL based spatial functions provided with the database should be preferred against using the QGIS based spatial functions from the API, although these are clearly provided in the QGIS menu bar and the toolbox and seem easier to use. Practical experience indicates that using the database functions yields more homogeneous software code and leverages the use of database indexing. Combination of data from multiple tables is much easier using SQL than writing explicit software code. Figure 1 demonstrates an SQL statement for connecting subcatchment areas to the pipe laying within the area:

```
UPDATE subcatchments
SET pipename =
( SELECT subcatchname
  FROM pipes
  WHERE within(centroid(subcatchments.geom),pipes.geom))
```

Figure 1: SQL statement for connecting pipes and subcatchments by location.

From the engineering point of view SQL encourages consulting engineers with a certain knowledge in programming languages to directly manipulate data using the SQL window in QGIS. Once a collection of approved SQL statements has been collected in a team of engineers it is easy to adapt an SQL statement to a similar use case. Spatial SQL commands are strongly standardized and can be used in combination with ordinary SQL statements, which after a certain training period leads to a very high productivity and flexibility.

An essential functionality is the possibility for the consulting engineer to work efficiently with the table data within the database. QGIS supports editing of table data with the easy-to-use QT Designer, which enables engineers to create straightforward forms including combo boxes displaying data from reference tables (Figure 2).

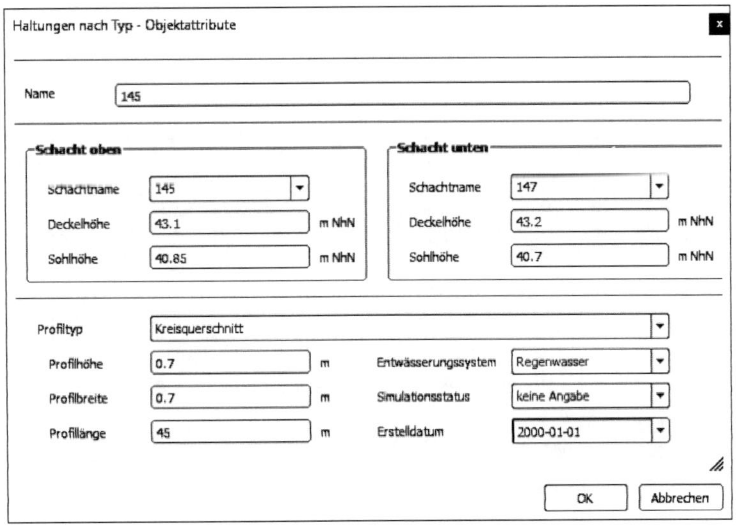

Figure 2: Pipe data editing form created with QT Designer.

4. Import and Export Functionality

There are three variants for exchanging data between a hydraulic simulation software and the QKan database. Some software programs offer a facility to copy paste the data, but it is useful primarily for small modifications by pasting the data into a text editor or a spreadsheet software (e.g. EXCEL). The second variant utilizes interchange formats defined by technical organizations or the software developer, which often implies a restriction on the extent of information. The third variant is the possibility of a direct access to the files or the database utilized by the simulation software.

4.1. Simulation software HYSTEM-EXTRAN

In Germany a widespread software (HYSTEM-EXTRAN, Institut für technisch-wissenschaft liche Hydrologie GmbH, Hannover, Germany) for hydrodynamic simulation of rainfall runoff and primarily originated from SWMM (EPA 2015), uses a Firebird database, which is accessible via Python requiring installation of the Firebird Server and the firebirdsql Python module. An import and an export plugin have been created (Höttges and Molitor 2016, Höttges 2017) covering all necessary drainage system data (pipes, nodes, pumps, weirs, pipe cross section data, catchment data, etc.) except rainfall data and simulation parameter files. The latter must be provided by the user by preparation of an empty database file created using the simulation software. Thus, the export module nearly allows an export-and-run-workflow because all data preparation, inspection and spatial data composition can be conducted using QKan/QGIS.

The QKan import plugin creates a SpatiaLite database and an appropriate project file (see below) from a Firebird database file containing the drainage system data. As an interesting secondary effect an authority responsible for the drainage system can visualize the results of a simulation project carried out by e.g. an engineering office without installation of the simulation software (Figure 3+4). This enables an engineering office to very easily exchange preliminary results to the customer throughout the design process and transfer the final results after completion of the project.

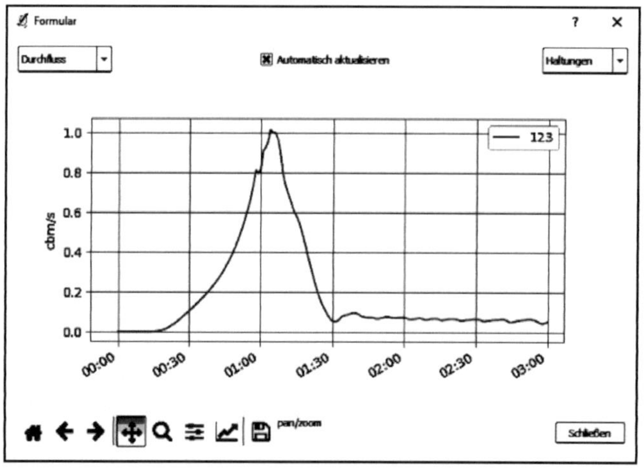

Figure 3: Discharge hydrograph.

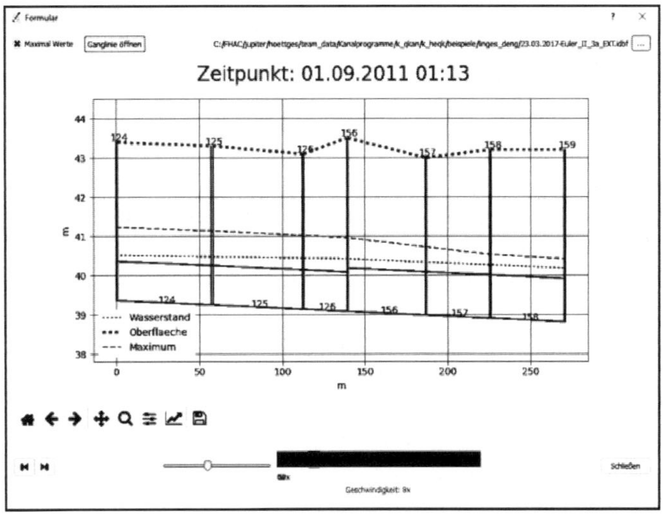

Figure 4: Longitudinal cross section with animation of water level.

5. Project File Template

Project files are a crucial point in all geographical information systems, because they hold all kind of graphical interpretation (layer style, thematic mapping, labeling, subject-specific symbols, etc.). They are enhanced continuously and are valuable as a template for future projects (Figure 5). In the case of sewage system data, point symbols show the type of drainage system nodes, line thickness represents the diameter of the pipes, line styles symbolize nodes connections by pumps and weirs. Furthermore, the import plugin analyses drainage system nodes on special criteria as peak points, sink points, start nodes or singular nodes to enhance the inspection of the data quality. All these definitions are stored in the project file.

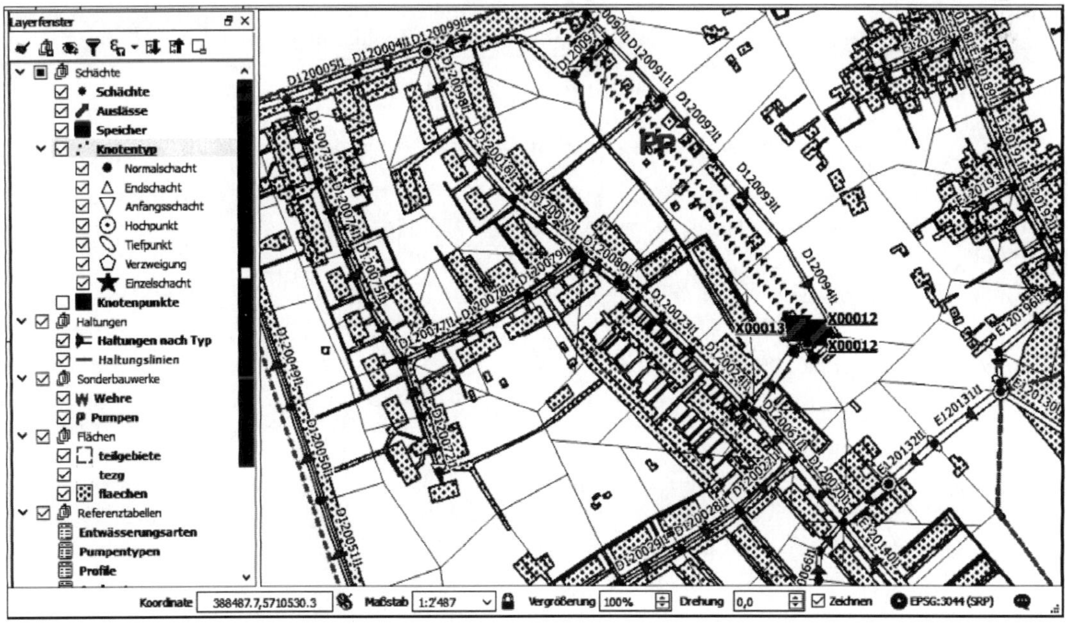

Figure 5: Sewage system plan after data import and customization of the template project file.

A template project file project.qgs is provided in the QGIS plugins folder. After completion of the data import the import plugin copies the template project file and treats it with an XML parser. Database path, coordinate reference system and display extent are adapted to the actual project, whereupon the project file is loaded into QGIS. The user can modify the project file and, if desired, use it as a new template by copying it to the template folder.

Furthermore, the user can establish his own set of templates. During the import it is possible to choose any other template project file as long as it includes layers connected to a QKan database. Additional layers remain unaffected.

The project file also contains the maps configured in the print composer, which is a very powerful tool for the consultant engineer for preparing printed layout plans. Very often title block templates exist, which have been designed with a CAD software. An efficient way to including such a file into a QGIS site plan is the free graphics software inkscape which imports Drawing Interchange File Format (DXF) and writes the Scalable Vector Graphics (SVG) format.

6. Further Development

The QKan project is still under development. The next plugins will enable data in- and export to the simulation software DYNA/Kanal++ (Tandler 2013). Additionally, a plugin for creating longitudinal sections in a CAD program will be realized with a direct link between the plugin and the CAD program.

References

EPA, 2015. Storm water management model users manual version 5.1. United States Environmental Protection Agency.

Fischer, K., 2015. Datenbankschema qgep. Online, online; accessed May 31, 2017.
URL `https://github.com/QGEP/datamodel/blob/master/diagram/datenbank_schema_von_koni_fischer.pdf`

Furieri, A., 2017. The gaia-sins federated projects home-page. Online, online; accessed May 31, 2017.
URL `http://www.gaia-gis.it/gaia-sins`

Höttges, J., 2017. Qkan import and export plugins for qgis. Online, online; accessed May 31, 2017.
URL `https://github.com/hoettges`

Höttges, J., Molitor, J., 2016. Qgis and database based system for managing urban drainage system data. Online, online; accessed May 31, 2017.
URL `https://av.tib.eu/media/20439`

Tandler, G., 2013. User manual. Online, online; accessed May 31, 2017.
URL `http://tandler.com/`

Processing Conservation Indicators with Open Source Tools: Lessons Learned from the Digital Observatory for Protected Areas

Lucy Bastin[a,b,*], Andrea Mandrici[a], Luca Battistella[a], Gregoire Dubois[a]

[a]*European Commission, Joint Research Centre (JRC), Directorate D: Sustainable Resources, Knowledge for Sustainable Development and Food Security, Via E. Fermi 2749, I-21027 Ispra (VA), Italy.*

[b]*Department of Engineering and Applied Science, Aston University, B4 7ET, Birmingham, UK*

Abstract: The European Commission has a commitment to open data and the support of open source software and standards. We present lessons learnt while populating and supporting the web and map services that underly the Joint Research Centre's Digital Observatory for Protected Areas. Challenges include: large datasets with highly complex geometries; topological inconsistencies, compounded by reprojection for equal-area calculations; multiple different representations of the same geographical entities, for example coastlines; licensing requirement to continuously update indicators to respond to monthly changes in the authoritative data. In order to compute and publish an array of indicators, we used a range of open source tools including GRASS, R, python, GDAL, PostGIS, geometry libraries for Hadoop, Geoserver, Geonode, and Mapserver. In addition we assessed the value of the commercial ArcGIS Pro software and the Google Earth Engine platform. We describe the lessons that we learnt in building and documenting a usable and repeatable workflow,highlighting weak spots and workarounds., Code for our processing workflows will be shared via github and key process flows will be shared via a VRE to allow reproducible research while complying with data redistribution restrictions from the data providers. Our final goal is to move the entire processing chain to open source tools and share it as a versioned resource.

*Corresponding author
Email address: `lucy.bastin@ec.europa.eu` (Lucy Bastin)

Submitted to FOSS4G 2017 Conference Proceedings, Boston, USA. *September 22, 2017*

1. Introduction

Assessing protected areas (PAs) for biodiversity conservation requires evaluation of characteristics such as the areas' connectivity and species assemblages (including the presence of threatened species), the uniqueness of their ecosystems, and the threats to which they are exposed. In this context, the Joint Research Centre of the European Commission has developed DOPA (the Digital Observatory for Protected Areas) a collection of Web tools supported by map, data and modelling services, designed to make biodiversity data readily accessible to policymakers, managers, researchers and other users (Dubois et al. 2016). The global indicators of DOPA support and frame the regional conservation picture for another set of tools developed by the JRC in the context of the Biodiversity and Protected Areas Management (BIOPAMA) programme, in particular the BIOPAMA Regional Reference Information System (http://www.biopama.org/observatories/access_the_rris). BIOPAMA builds capacity in Africa, Caribbean and Pacific to improve decision-making related to PA management. Given the importance of these decisions, the currency, consistency and quality of the data produced and shared by JRC is key in order to maintain a reliable service which is comprehensible and useful to a variety of stakeholders and decision makers.

As the technical staff responsible for populating the databases and Web services of DOPA, we realized that other individuals and institutions are undoubtedly facing the same challenges as us, with the same datasets. We have summarized some elements of our story in this manuscript, in order to start a conversation about data sharing and processing which we hope can lead to more effective collaborations in the future.

2. Our Technical Priorities

2.1. Robustness and transparency of methods

DOPA uses peer-reviewed methods to calculate a range of indicators on PA state, pressures and threats (Dubois et al. 2016); for example:

(1) A species irreplaceability index (SII) is computed according to the method of Le Saout et al. 2013 which scores each PA according to its relative coverage of suitable range for selected species. The SII score may relate to a specific taxon and/or threat group, or an overall score; it does not incorporate any consideration of complementarity in the network.

(2) The connectivity indicator developed by Saura et al. 2017 allows a user to distinguish between the proportion of a country or ecoregion which is formally protected and the amount of that protection which is adequately connected to allow movement of animals across the landscape, considering a range of different dispersal capacities.

2.2. Reliability and openness of datasets and workflows

Our goal is to use openly available datasets with clearly documented provenance, including some high-quality datasets developed in-house at JRC. The recently-published Global Surface Water product (Pekel et al. 2016) was analyzed to record area and net 32-year change of permanent / seasonal water within each PA. To evaluate human pressures on PAs, we use the Global Human Settlement map (Pesaresi et al. 2016), and in future we hope to exploit the outputs of projects such as ROADLESS (http://forobs.jrc.ec.europa.eu/roadless/) which detect and delineate the unofficial logging roads which can lead to further forest encroachment. All DOPA indicators are made available using open OGC standards and publicly-accessible REST services (http://dopa-services.jrc.ec.europa.eu/services), and can be consumed by a wide range of web clients such as DOPA Explorer (http://dopa-explorer.jrc.ec.europa.eu/dopa_explorer/). The European Commission has a commitment to open data and open source software. Wherever possible, we use free and open source tools and support the use of these tools by developing and sharing code in GitHub (https://github.com/doctorluz/py-utils/blob/master/PostGISTableToAvro.py).

3. Challenges

DOPA analyses and aggregates global datasets encapsulating highly complex phenomena, which naturally contain certain generalizations and omissions. Issues such as data accuracy and completeness are outside the scope of this paper, since we assume that these datasets are the best possible representations of the global state of protection and of potential species ranges. Figure 1 shows the post-processing steps for one example dataset (UNEP-WCMC's World Database on Protected Areas (WDPA) (https://www.protectedplanet.net/)), to provide some context for the discussions below.

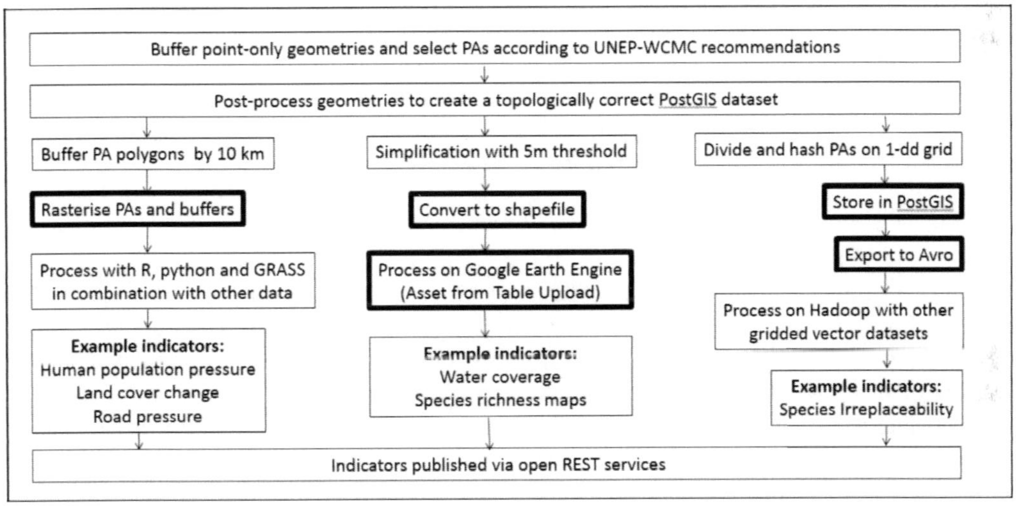

Figure 1: Processing steps for the protected area polygons. Boxes in bold indicate value-added datasets which, legal agreements permitting, could be usefully shared with other researchers to ensure consistency and reduce duplicated effort.

3.1. Staying up-to-date

The WDPA, documenting over 200,000 PAs globally, is updated every month, and information is published on the countries where PAs have been updated (https://www.protectedplanet.net/c/monthly-updates/2016/november-update-of-the-wdpa). The ideal would be to compute new metrics only for those PAs which have been added or altered, except in the case of complementarity-based indicators where the whole set of PAs will need to be reprocessed. However, at the time of writing, it is not possible to obtain a 'change-only' update. An example of good practice on this topic is the GAUL dataset of administrative boundaries, where changes are tracked using unique IDs (http://www.fao.org/geonetwork/srv/en/metadata.show?currTab=simpleid=12691).

3.2. Legal restrictions

While the datasets used for DOPA are open for public use, redistribution of data and derived products is more tightly controlled. For example, PAs can be displayed using our Mapserver or Geoserver Web Map Services, but not exposed via a Web Feature Service; and currently, we cannot share post-processed data with partners in order to ensure consistency. Since many analysts around the globe are using these datasets, it is certain that much data preparation work is being duplicated, but the curating agencies do not have the resource or mandate for this post-processing. In some cases, generation of derived products requires prior approval, a restriction which limits scientific research by third parties. Our current approach to this challenge is described in section 5.1.

3.3. Different representations of the same thing: the example of coastlines

For DOPA processing, we use datasets representing the countries of the world (GAUL, FAO https://www.protectedplanet.net/), Exclusive Economic Zones (http://www.marineregions.org /eezsearch.php), terrestrial ecoregions (Olson et al. 2001) (https://www.worldwildlife.org/biomes), marine and pelagic ecoregions (https://www.worldwildlife.org/publications/marine-ecoregions-of-the-world-a-bioregionalization-of-coastal-and-shelf-areas), coastal protected areas, and species ranges intended to delineate terrestrial or marine specialization. These coastline representations rarely or never match, leading to numerous sliver polygons, and to apparent inaccuracies such as narwhals being found on land. Nevertheless, it is crucial to derive a harmonized and consistent base data layer, since DOPA has important responsibilities in supporting regular reporting initiatives such as the Protected Planet report (UNEP-WCMC and IUCN 2016) and the country reports (https://www.cbd.int/reports/) for the Convention on Biological Diversity. The CBD is currently using the indicators provided through the DOPA Explorer to support the preparation of country dossiers for Aichi Targets 11 and 12 (https://www.cbd.int/sp/targets/), and has also encouraged its Parties to consult DOPA in the revision of their NBSAPs (National Biodiversity Strategies and Action Plans https://www.cbd.int/nbsap/). Typical metrics which must be computed in this context are: the area of an ecoregion which falls under a country's responsibility; the amount of protection for each terrestrial or marine ecoregion within a country; the relative contribution that a country is making to the protection of an ecoregion worldwide; and the number of different ecoregions which fall within a particular protected area. The last metric gives a very generalized estimate of habitat diversity, which is then supplemented by detailed biophysical modelling and assessment of marine floor geomorphological features. However, even the relatively simple overlay of countries, EEZs, ecoregions and protected areas required for the generalized estimates can be computationally demanding. For purposes of transparency, reproducibility and data consistency, it is necessary to build, from available open data sources, a high-quality, topologically correct base layer encapsulating legal country boundaries permissible for use by the European Commission, EEZs and ecoregions. The complex procedure, which currently involves the use of a wide range of software to handle the different steps, is described in Section 5.2.

3.4. Technical and interoperability challenges

The datasets that we integrate use a variety of data models and standards. For example, the topological models of PostGIS and ESRI differ in terms of legal coordinate ordering for internal rings, and ESRI shapefiles, which are frequently used as an exchange format, comply with a Simple Feature data model rather than the Topological model required by GRASS. Reprojection to equal-area projections such as Mollweide to assess real-world extent can introduce self-intersections, even where polygons were originally topologically correct. Some datasets span the range from -180 to 180 decimal degrees, while others (e.g. http://www.marineregions.org/download_file.php?fn=v9_20161021_HR_0_360) can be obtained in a 0-360 degree version.

Even when a workflow is documented, operations are often specified without details of the specific software or algorithm used. For example, geodesic buffers are recommended in the processing guidance for the WDPA (http://wdpa.s3.amazonaws.com/WDPA_Manual/English/WDPA_Manual_1_4_EN_FINAL.pdf) and are also used in DOPA to compare conditions inside and outside PAs. However, different software can yield very different results. For PAs which cross the dateline, ArcGIS produces a geometrically-correct buffer in two segments, while both PostGIS and the Hadoop spatial framework (https://github.com/Esri/spatial-framework-for-hadoop) create a self-intersecting polygon that spans the entire globe. Thus, in order to ensure consistency, we translate such polygons to a 'safe' longitude before buffering them, then translate the buffer back to its original location and, if necessary, split the buffer on the dateline. The whole workflow, including this extra step, is described in an open access document (https://dopa.wikispaces.com/WDPA+protected+Area+boundaries), and summarized in Section 5.1.

3.5. 'Big Data' problems

Given the global nature of DOPA and its ambitious aims, the input datasets are necessarily large and extremely topologically complex. For a subset of our metrics, only PAs over 50 square km are processed, (this amounts to 98% of global protected area) but even this reduced dataset has over 20,000 polygons. The most complex PA (Vindellven in Sweden) has over 862 thousand vertices, and the species ranges for the IUCN Red List are equally challenging, particularly for many bird species and marine mammals, which span the globe while incorporating complex coastlines. Rasterization of the data simplifies computation, but also brings new challenges and tradeoffs: for example, when calculating zonal statistics for large regions and small (<=30m) pixels, on a distributed platform such as Hadoop or Google Earth Engine, tasks must be carefully partitioned across space and results re-aggregated from smaller regions, to minimise communication between multiple nodes. Large PAs close to the poles are especially problematic, especially if processed in a non-sinusoidal projection, since the number of pixels to be aggregated rises far beyond the actual area for which information is being derived. Vector processing and data sharing with Google Earth Engine is also fast advancing: Fusion Tables could each contain a maximum of 1 million vertices, but have been very recently superseded by the more flexible Table Upload tool (https://developers.google.com/earth-engine/importing).

4. Potential Solutions

As a platform for storing, cleaning, processing and managing vector data, and for underpinning the REST service interface that delivers DOPA indicators to Web clients, PostGIS (http://www.postgis.net/) has been crucial to our work. At the time of writing, we are setting up a more powerful instance of PostGIS 9.6, which will allow us to exploit multi-threading capabilities and evaluate where PostGIS Raster can be of value (for example, speeding up overlay of relatively small raster polygons on tiled global datasets). Other tools such as GRASS (https://grass.osgeo.org/), with its well-established functions for vector and raster processing, and python and R spatial libraries, have been heavily used, along with fundamental libraries such as gdal (http://www.gdal.org/). For certain vector analyses we have made use of Hive on a Hadoop cluster (see section 5.3) and for pixel-wise computations requiring access to large archives of EO data or global derived products, and for easy prototyping of the MapReduce approach, we have benefitted from the capabilities of Google Earth Engine (https://earthengine.google.com). As an organization with heavy network security, it is difficult to install specific modules and libraries on shared processing servers, and docker (https://www.docker.com) has been extremely useful as a means to set up and share self-contained working environments with specific collections of software and models.

We also exploit established best practice in the geospatial field, by partitioning the data in a smart way so that it is more tractable for processing while also being as reusable as possible. For relatively static datasets we tested data-dependent griddings, which as closely as possible equalized vertex number in cells of varying resolution. However, a common grid is required for more dynamic datasets and for sharing with other researchers. Currently we have settled on a 1-decimal degree tiling (described in Section 5.3) but equal-area grids such as the ISEA hexagonal grid supplied in IUCN's Red List Toolbox (http://www.iucnredlist.org/technical-documents/red-list-training/iucnspatialresources) are also good candidates. In many cases, this is a matter more of cultural than of technical interoperability. Given the good progress of OGC's Discrete Global Grid Standards Working Group towards an accepted standard and implementation pilots, we will follow the developments and demonstrators of this community, in order to be ready to adopt useful tools and data structures which emerge to support interoperability.

5. Example Case Studies

5.1. Sharing a processing workflow: Preparing the WDPA for use on several platforms

Our processing workflows (e.g. Figure 1) are described in open-access documents (http://www.marineregions.org/eezsearch.php), but, as described above, this is not sufficient to guarantee a reproducible result. Therefore the code will also be shared on GitHub and published as a Data Paper. The long-term goal is to share the processing itself through a Virtual Research Environment. This processing workflow can now be almost entirely executed using PostGIS (steps 1-8) and ogr2ogr (steps 12-13) after the initial import of the provider's feature geodatabase or shapefile (Bastin and Mandrici 2017). The one section of the workflow which remains intractable using PostGIS, ArcGIS Pro or GRASS is the generation of external buffers for every PA (steps 8-11). This necessitates the use of Hadoop for this step at present, but the first test for our emerging, more powerful database instance which will exploit the multi-threading capabilities of PostGIS 9.6, will be to evaluate whether this step can be brought back into the PostGIS processing chain.

The relevant WDPA version (IUCN and UNEP-WCMC 2017) is downloaded in a lat/long EPSG:4326 projection, and processed as follows:

(1) Protected areas with point-only geometries and a reported area >0 are given a circular geodesic point buffer with the reported area.

(2) Polygon self-intersections are corrected using the ST_MakeValid function of PostGIS, and point and line geometries resulting from the correction are discarded.

(3) Polygons digitized with only two coordinates (these mainly consist of sunken US ships which provide useful marine habitat) are converted to lines and buffered by a very small distance to create valid polygons.

(4) Features at the dateline whose geometries fall outside legal coordinates are split, and the resulting part of the geometry are each translated to the correct side of the dateline.

(5) Where a PA is divided into several parcels representing zones (separate records in the WDPA), these are aggregated to form a single union, so that WDPA ID is the unique identifier for the dataset. Any internal slivers and topological inconsistencies which result are repaired and removed. In these cases, (around 450 of the >200,00 PAs) parcel names, IDs etc are concatenated in a comma-separated string to allow a single record for each PA.

(6) *(January 2017 version only), to deal with erroneous duplicate PA IDs in the original dataset, the ID for Polesye Valley of River Bug, Belarus should be changed to 555624313 (the ID value for March 2017).*

(7) As recommended by WCMC (http://wdpa.s3.amazonaws.com/WDPA_Manual/English/WDPA_Manual_1_4_EN_FINAL.pdf), the data is filtered to remove all features with a status of "not reported" or "proposed", and all features designated as UNESCO Man and Biosphere Reserves.

For a subset of protected areas exceeding 50 square km in area, 10-kilometre buffers are produced in order to compare conditions inside and outside the PA, as follows:

(8) Polygons where any part of the PA or its buffer would cross the dateline are translated horizontally by 180 degrees to ensure correct buffering (This is because of issues with many buffering functions at the dateline, including the ESRI libraries available for use on Hadoop).

(9) All polygons are processed using Hive on a Hadoop cluster to produce an external geodesic buffer of 10 km.

(10) Polygons which were translated in step 8, along with their buffers, are split as appropriate and moved back to their correct positions on the dateline.

(11) Buffer areas are masked to eliminate areas which, although in the exterior buffer of a protected area, are covered by protection from another PA - i.e., to include only unprotected regions.

Additional processing for raster products

(12) The buffered PAs are rasterised individually to produce geotiffs with a background value of 0, a pixel value of 1 in the buffer and 2 in the PA itself. The resolution of these rasters is c. 100m at the equator, in a projection of EPSG:4326.

(13) For masking purposes (see step 11) and for use on Google Earth Engine for computing country-level protection of water and forest, a binary raster containing all PAs with polygon geometries is generated at a matching resolution to step 12.

The code for all above steps will be shared on github, and wrapped as python or R modules to allow easier configuration and progress tracking. The latter implementation will be hosted on a VRE by the end of 2017, to allow other users and research partners to generate and document data consistent with our processing methods. *For information on the progress of this initiative, please contact the authors on the email addresses provided.*

5.2. Building a consistent, well-documented base layer: Countries, EEZs and ecoregions

DOPA disseminates indicators based on intersections of the WDPA dataset with other data including population, landcover, etc. However, in order to correctly attribute protection efforts to the relevant countries and governments, and to assess the coverage of the world's habitats by protection, a reliable base layer encapsulating the combination of political and habitat boundaries is required. The generation of a consistent dataset which covers all areas of the globe with no overlaps has been challenging, and is documented below. Where possible, open source solutions have been used, but a few steps have required the memory management of ArcGIS

Pro. We are actively seeking means to move these steps into the open-source domain in order to avoid restrictions on reproducibility.

5.2.1. Country boundaries

This dataset is built from a combination of GAUL country boundaries and EEZ exclusive economic zones (referenced earlier). The two datasets come with different coastlines, different attribute structures; (they have only one common field: ISO3), and small differences within the content (for example, sovereignty updates).

Normalization of these two datasets involved (in brackets are the tools used at each stage):

(1) Intersection: where there is overlap, GAUL wins (QGIS, PostGIS, ArcGIS Pro).
(2) Gap filling: where there is a gap between the two coastline, this is filled and attributed to the containing EEZ (ArcMap 10.3).
(3) Unification of the attribute structure (PostGIS).
(4) Identification of Disputed, Joint Managed areas, and attribution to multiple ISO3 codes (PostgreSQL).
(5) Identification of the areas not covered by any of the two dataset as ABNJ - Area Beyond National Jurisdiction (SAGA).
(6) Identification of updated information according to the ISO 3166-1 standard (PostgreSQL).
(7) Dissolve to remove non-unique keys, where present (PostGIS).
(8) Geometric simplification (10 meters, preserving topology) (PostGIS).
(9) JOIN with ancillary information (ISO2 and ISO numeric/UN_M49) to include additional codes.
(10) JOIN with ancillary information to get country grouping information for example, continental and regional groupings which are of relevance for conservation funding and policy decisions (PostgreSQL).

The result is a dataset (Figure 2, left) with updated geometries and structure, which maintains the original source codes (ADM0 for GAUL, MRGID for EEZ).

5.2.2. Ecoregion boundaries

Sources for this dataset are:

(1) **Terrestrial Ecoregions of the World**: biogeographic regionalization of the Earth's terrestrial biodiversity. (TEOW) (Olson et al. 2001).
(2) **Marine Ecoregions and Pelagic Provinces of the World**: biogeographic classification of the world's coastal, continental shelf, and surface pelagic waters (The Nature Conservancy (TNC) 2012). This dataset combines two separately published datasets:
 (a) **Marine Ecoregions Of the World** (MEOW) (Spalding et al. 2007) and
 (b) **Pelagic Provinces Of the World** (PPOW) (Spalding et al. 2012)

The two datasets differ in their attribute structure. Normalization involved:

(1) Intersection: where there is overlap (MEOW/PPOW comes without coastline), TEOW wins (ArcGIS Pro).
(2) Unification of the attribute structure (PostgreSQL).

The result is a tessellation of the world with no gaps, where every polygon can be allocated to a terrestrial or marine ecoregion, or a pelagic province (Figure 2, right).

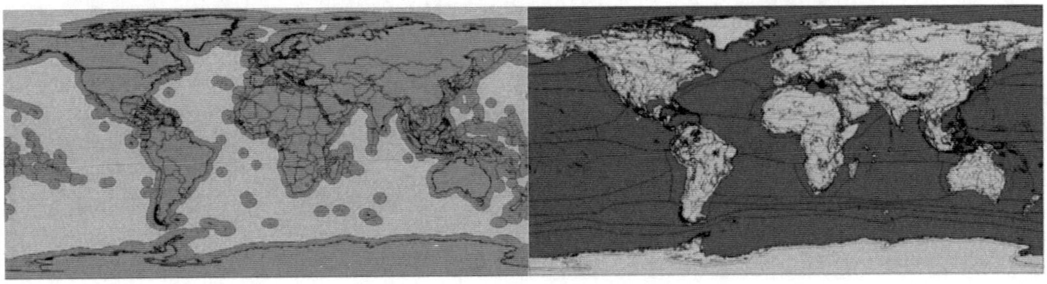

Figure 2: The final base layers for countries/EEZs (left) and ecoregions (right).

5.2.3. Unified base layer

To reduce the number of processing loops in computing indicators for the WDPA, the two above datasets are merged. The target is to achieve, with a single processing cycle, information on:

(1) Country
(2) Ecoregion
(3) Terrestrial/marine breakdown of the country and many further statistics related to the above as:
- (a) number of PAs per country
- (b) number of marine/terrestrial/coastal PAs per Country
- (c) area/percentage of a country's ecoregions which are protected
- (d) number of PAs per ecoregion
- (e) etc.

Merging of the two datasets involves:

(1) Dissolve of the country/EEZ on country codes into larger polygons bounded by the country's EEZ extent. In other words, the coastlines visible in Figure 2 are discarded. (PostgreSQL, ArcGIS Pro).

(2) Intersect (UNION) of the geometries and attributes of the result from the above step, and ecoregions. This means that, for the purposes of this analysis, the junction between land and sea is determined by the TEOW dataset rather than the GAUL geometries (ArcGIS Pro). The consequences of this decision are illustrated in Figure 3, and further discussed below.

Each polygon contained in the resulting dataset contains information on ecoregions (TEOW, MEOW and PPOW original codes), country (ISO3/ISO2/UN_M49), and presence of land or sea (as defined by the TEOW coastline, which is lower resolution than the GAUL coastline, but is still adequate to this particular analysis, given the data resolutions of the phenomena that are being summarized at a global scale).

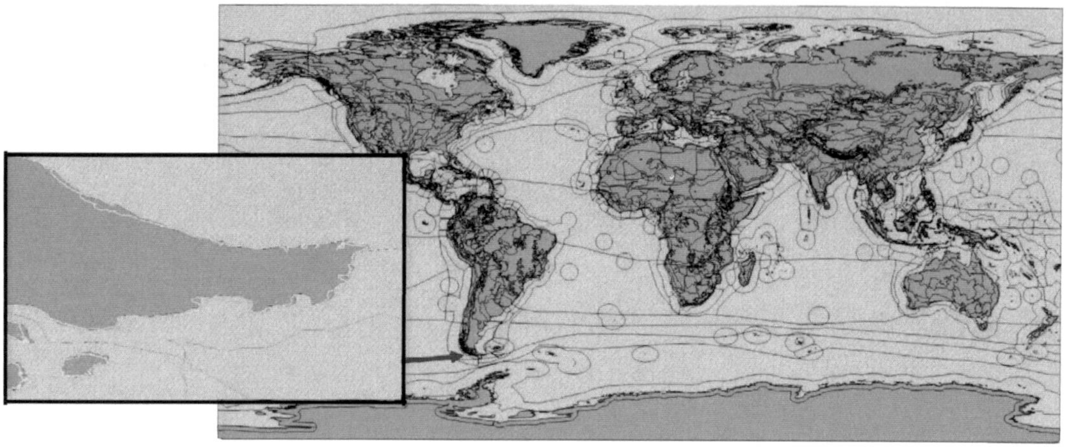

Figure 3: Unified Base layer, with example of discrepancy between GAUL (yellow) and TEOW (black) coastlines.

5.3. Multi-way intersections species ranges and protected areas

The SII described in section 2.1 requires a demanding multi-way intersection between over 200,000 protected areas and over 27,000 species ranges. This operation is achieved by a segmentation of the data on a one-decimal degree grid to form a list of either simple squares (where the grid cell is completely contained) or smaller polygons. When applied to all input datasets, this approach allows a straightforward SQL query using the Hadoop ESRI spatial framework, where an ST_INTERSECTION operation is only necessary for those cells which are not completely contained by either polygon. A bespoke Java class written to support this procedure can be found on GitHub (Lars Francke, 2014 https://github.com/lfrancke/jrc/blob/master/src/main/java/jrc/CellCalculator.java).

Even when the task is thus simplified, there are still occasional errors in the first (union) step of the procedure, and these are much harder to debug once the data is distributed across the Hadoop cluster. Therefore the gridding and union steps were moved off the cluster, as shown in Figure 1. This has the added benefit of making helpfully-gridded and indexed data available for use on other platforms.

6. Lessons Learnt

The wide range of demanding analyses behind DOPA mean that different tools are suited to each step of the workflow, and this has been challenging in terms of provenance documentation and data models / formats. However, careful preparation of the data can make it more easily re-usable across a variety of platforms. We have learnt that at present, some steps of our processing still need to be carried out using commercial software (for example, large-scale intersections in ArcGIS Pro), but are very close to testing more powerful and better-supported instances of open source alternatives on JRC's new JEODPP Big Data infrastructure.

Most users of environmental datasets are trying to do reproducible and accountable science, but different post-processing workarounds and tools can lead to published results which are not repeatable or comparable. To work more effectively, we would ideally share value-added data processed to an agreed standard and format. Since legal restrictions currently forbid this type of redistribution, the next best solution is to share the processing workflow, including code and environmental settings or parameters. Either the whole environment can be shared (for example using docker or vagrant) or the processing itself can be published as a service (for example, DOPA's marine geomorphological metrics are computed in collaboration with the BlueBridge (https://bluebridge.d4science.org/explore) project, which uses the D4Science VRE infrastructure).

Processing power alone has not helped us to break down and overcome these challenges: far more fruitful has been the process of thinking through and classifying the problems faced, and we hope to share this active reflection process with other researchers in future.

References

Bastin, L., Mandrici, A., 2017. WDPA processing workflow description. Online; accessed May 31, 2017. URL `https://dopa.wikispaces.com/WDPA+protected+Area+boundaries`

Dubois, G., Bastin, L., Bertzky, B., Mandrici, A., Conti, M., Saura, S., Cottam, A., Battistella, L., Martinez-Lopez, J., Boni, M., 2016. Integrating multiple spatial datasets to assess protected areas: Lessons learnt from the digital observatory for protected areas (dopa). ISPRS International Journal of Geo-Information 5, 242.

IUCN, UNEP-WCMC, 2017. The World Database on Protected Areas (WDPA). Cambridge, UK: UNEP-WCMC. Online; accessed May 31, 2017. URL `www.protectedplanet.net`

Olson, D. M., Dinerstein, E., Wikramanayake, E. D., Burgess, N. D., Powell, G. V. N., Underwood, E. C., D'Amico, J. A., Itoua, I., Strand, H. E., Morrison, J. C., Loucks, C. J., Allnutt, T. F., Ricketts, T. H., Kura, Y., Lamoreux, J. F., Wettengel, W. W., Hedao, P., Kassem, K. R., 2001. Terrestrial ecoregions of the world: a new map of life on earth. Bioscience 51(11), 933–938.

Pekel, J. F., Cottam, A., Gorelick, N., Belward, A. S., 2016. High-resolution mapping of global surface water and its long-term changes. Nature 540, 418422.

Pesaresi, M., Ehrlich, D., Ferri, S., Florczyk, A. J., Freire, S., Halkia, S., Julea, A. M., Kemper, T., Soille, P., Syrris, V., 2016. Operating procedure for the production of the Global Human Settlement Layer from landsat data of the epochs 1975, 1990, 2000, and 2014. Publications Office of the European Union, EUR 27741 EN.

Saout, S. L., Hoffmann, M., Shi, Y., Hughes, A., Bernard, C., Brooks, T. M., Bertzky, B., Butchart, S. H. M., Stuart, S. N., Badman, T., Rodrigues, A. S. L., 2013. Protected areas and effective biodiversity conservation. Science 342, 803–805.

Saura, S., Bastin, L., Battistella, L., Mandrici, A., Dubois, G., 2017. Protected areas in the worlds ecoregions: how well connected are they? Ecological Indicators 76, 144–158.

Spalding, M. D., Agostini, V. N., Rice, J., Grant, S. M., 2012. Pelagic provinces of the world): a biogeographic classification of the worlds surface pelagic waters. Ocean and Coastal Management 60, 19–30.

Spalding, M. D., Fox, H. E., Allen, G. R., Davidson, N., 2007. Marine ecoregions of the world: A bioregionalization of coastal and shelf areas. Bioscience 57, 573583.

The Nature Conservancy (TNC), 2012. Marine Ecoregions and Pelagic Provinces of the World. GIS layers developed by The Nature Conservancy with multiple partners, combined from Spalding et al. (2007) and Spalding et al. (2012). Cambridge (UK): The Nature Conservancy. Online; accessed May 31, 2017. URL `http://data.unep-wcmc.org/datasets/38`

UNEP-WCMC, IUCN, 2016. Protected planet report 2016. UNEP-WCMC and IUCN: Cambridge UK and Gland, Switzerland.

The Billion Object Platform: A System to Lower Barriers to Big, Streaming, Spatio-Temporal Data Sources

Devika Kakkar[a,*], Ben Lewis[a], David Smiley[b], Ariel Nunez[c]

[a] *Center for Geographic Analysis, Harvard University 1737 Cambridge Street, Cambridge, MA, USA, 02138*

[b] *Self Employed, Freelance Consultant Cambridge, MA, USA*

[c] *Terranodo LLC San Diego, CA, USA*

Abstract: With funding from the Sloan Foundation, Boston Area Research Initiative (BARI), and Harvard Dataverse, the Harvard Center for Geographic Analysis (CGA) has developed a big spatio-temporal data visualization platform called the Billion Object Platform or "BOP". The goal of the project is to lower barriers for scholars who wish to access large, streaming, spatio-temporal datasets. Since once archived, streaming data gets big fast, and since most GIS systems don't support interactive visualization of millions of objects, a new platform is needed. Our instance of the BOP is loaded with the latest billion geo-tweets and is fed a real-time stream of about 1 million tweets per day. The CGA has been harvesting and archiving geo-tweets since 2012. Tweets flowing into the BOP are enriched with sentiment and census information to support further analysis. Incoming and intermediate data is streamed/stored in Apache Kafka. The core of the BOP is Apache Solr, which supports fast search. Some significant Solr enhancements were developed (and contributed back) – notably 2D "heatmap faceting" to support spatial visualization. The BOP fronts Solr with a RESTful web service, which provides a friendly, and secure API that is accessed from a browser-based client. The client dynamically displays temporal and spatial distributions of results for result sets containing hundreds of millions of features. The system is open source and runs on commodity hardware. The geo-tweet instance is hosted on Massachusetts Open Cloud (MOC), an OpenStack environment (OpenStack 2017). All components are deployed in Docker and orchestrated by Kontena.

*Corresponding author
Email address: kakkar@fas.harvard.edu (Devika Kakkar)

1. Introduction

1.1. Background

The Billion Object Platform (BOP) is a prototype platform designed to lower barriers for researchers who need to access big streaming spatio-temporal datasets. Most mapping systems today are extremely constrained when it comes to representing millions of features on a map unless the visualization has been previously generated (Lewis 2016a). The system provides an approach for managing real-time streaming flows of big data, and presents a prototype for storing, indexing, visualizing, and querying geo-tweets which are tweets containing a GPS coordinate from the originating device.

The system exposes the latest billion georeferenced tweets geo-tweets for exploration by the basic dimensions of time, space, keyword, and the system supports interactive visualization of distributions of tweets in both time and space. Approximately, 1-2% of tweets are geo-tweets and Centre for Geographic Analysis (CGA) at Harvard University has been harvesting them since 2012, creating an archive of about 8 billion objects to date. The BOP further enhances 2D faceting (heatmap) capabilities in Lucene and Solr that were originally developed by the CGA in a previous project to build a map services search platform called HHypermap Registry which was funded by the National Endowment for the Humanities (Lewis 2016b).

1.2. Functional Requirements

The goal of the BOP is to make it easier for a scholar to explore a big dataset and create the right subset for their research purposes. The functional requirements of the system were designed to push the envelope of what is possible in geo-visualization:

- Provide access to the most recent billion geo-tweets
- Support real-time search
- Sub-second queries including heatmaps and temporal histograms
- Implementable on the cheap using commodity servers

In addition to attempting to address a general big data visualization problem, the BOP is configured to operate within the context of a particular data archive, Harvard Dataverse, exposing an API that can be plugged into Dataverse to enable subsets of the BOP to be extracted, then pushed into Dataverse for further analysis. This instance of the BOP is loaded with tweets but a similar dataset could, with some tweaking, be substituted.

1.3. System Architecture

The high-level architecture of the BOP is shown in Figure 1 (Smiley 2016). The primary infrastructure components are Apache Kafka and Apache Solr. Tweets are harvested and then enriched with metadata in the form of sentiment as well as spatial joins with census and other canonical boundary codes. The geo-tweets are archived using a long-term Kafka topic. The BOP itself, which consists of a Solr index based copy of the data, represents the latest billion while the index goes back further in time. The BOP-core exposes a search and extraction web service API that the client consumes. All components are deployed to a Docker based infrastructure managed by Kontena. The term BOP-core is used to refer to both the Solr index and the web service that exposes it.

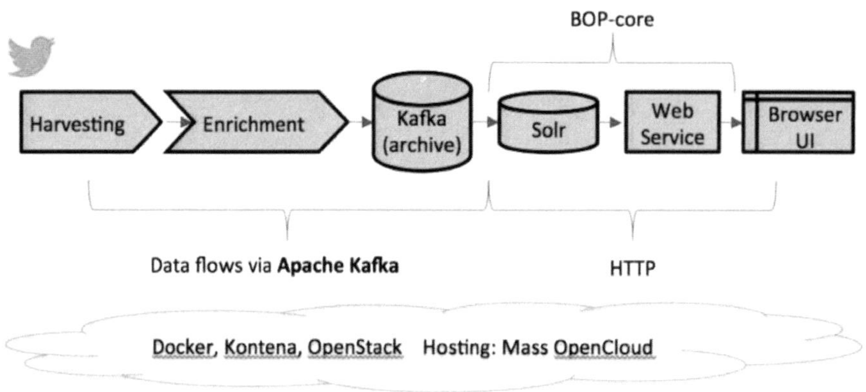

Figure 1: Logical high-level architecture of BOP.

1.4. Preliminary Performance Results

The performance of the system is difficult to characterize, as there are many variables at play including: the document size of the largest Solr shard being queried, the number of docs matching a query in that shard, whether the same query has been run before (thus cached), the nature of the filters (keyword, space, time, others), and the optional components of a query: requesting top docs, requesting heatmaps (at various resolutions), requesting temporal histograms (at various resolutions), and whether or not the leading shard (for the current time period) is queried.

The table of times below has two columns; both were for separate 12-month periods (across 12 Solr shards). More data exists in the repository for 2015 than 2016. Times are in milliseconds.

Table 1. Performance Results.

	Largest Shard	
	2016 (17.8M docs)	2015 (139M docs)
Filters (F)	1100 ms	3110 ms
F and top docs	800 ms	930 ms
F and heatmap	65 ms	365 ms
F and histogram	22 ms	25 ms

Table 1: Performance Results.

The first row is the cold (non-cached) time it took to return a simple document count for the keyword query usa along with the temporal period for that year, and a spatial filter covering the USA, northeast region. The temporal and spatial aspects were cached but the keyword was chosen anew. Performance can vary based on a number of factors. If the query has been issued before, performance is much faster. The second row is the cold (non-cached) time for returning the 25 top docs (sorted by time). Again performance can vary. If the query has been previously issued, this time is drastically reduced. The third row is the time to compute a heatmap of 6003 cells. This is not sensitive to caching because it is not cached. The fourth row is the time range for the temporal histogram: 53 weeks. The performance of this and the sensitivity to caching depends on the presence of filters. It can be much slower for queries lacking a keyword filter.

2. Components

2.1. Harvesting/Archiving

Harvesting is the process of capturing real-time data and loading it to the BOP. A harvesting service continuously polls Twitter's API based on predefined users and coordinate extents, then deposits the resulting JSON to a Kafka topic after serializing the tweet JSON using MessagePack. The latest billion geo-tweets are then pushed to Solr and made available for fast search and exploration via the BOP API (API 2017). CGA has been harvesting tweets since 2012 and has gathered about 8 billion objects to date.

2.1.1. Removal of Selected Bots

A twitter bot is a software program that sends out automated posts with auto-generated (often seemingly random) geospatial coordinates via Twitter. These geo-tweets, usually spread arbitrarily across the globe, create static in the geo-visualization without providing much benefit. We remove such bots at harvest time by user name, as we identify them. We document the bots we have removed so users can still retrieve them if needed directly from the Twitter API.

2.2. Enrichment

Enrichment consists of binary sentiment analysis (e.g. happy/sad) and geo-enrichment, e.g. census code, by reverse-geocoding against census and other boundary types. Both enrichment types will be discussed in detail below. As seen in Figure 2, this service consumes an input Kafka topic, and then adds the enriched tweets to another topic.

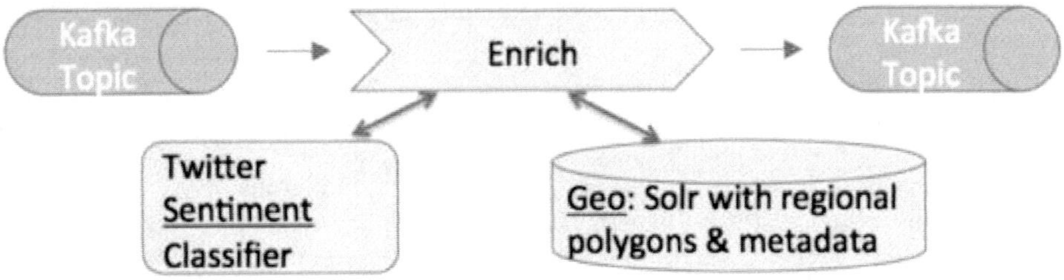

Figure 2: Process of enrichment in BOP.

2.2.1. Sentiment Analysis

Sentiment analysis is a field of study which attempts to identify the emotion, expressed in text using Natural Language Processing (NLP) tools. Social media platforms such as Twitter provide a constant source of textual data, much of which is expressed with an emotion, which can be determined using Sentiment Analysis tools. We are using the Support Vector Machine (SVM) algorithm with a linear kernel, which is written in python and uses the scikit-learn library. Two classes of sentiment are assigned: positive (1) and negative (0). Tweets are pre-processed before sentiment stamping. The training corpi used are Stanford's Sentiment140, Polarity dataset v2.0, and the University of Michigan's SI650-sentiment classification dataset.

The process of sentiment analysis is divided in two phases: training and prediction as shown in Figure 3. In the Training phase we train the classifier and save it as a pickle file. In the prediction phase we load the trained classifier. Thereafter, for each tweet, we parse, pre- process, stem, and predict the sentiment. The precision, recall and F1-score of the algorithm is 82% for each. Sentiment enrichment takes an average of 20 milliseconds per tweet with no emoticon and 5 milliseconds for a tweet with an emoticon.

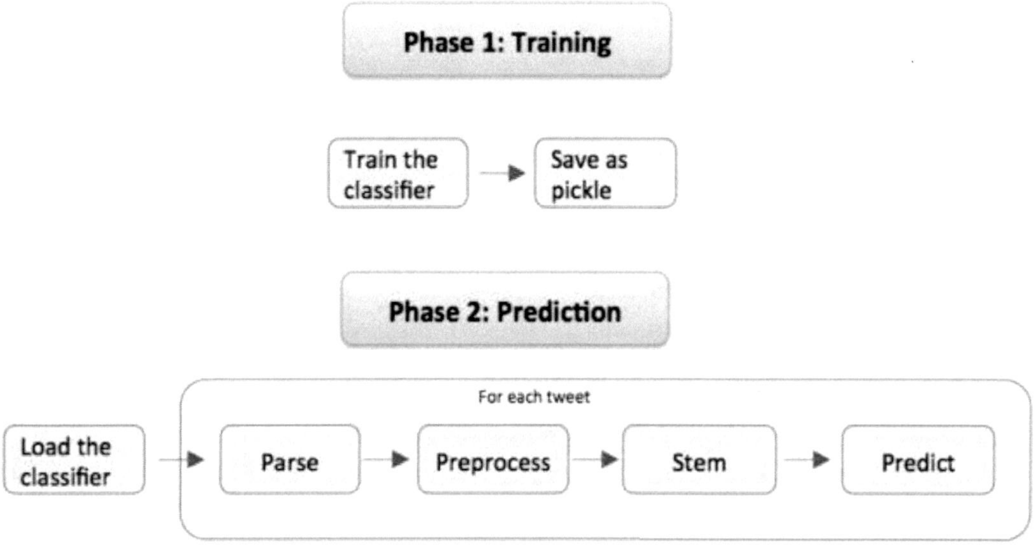

Figure 3: Sentiment analysis in BOP.

2.2.2. Solr for Geo-enrichment

Geo-enrichment uses Reverse Geocoding to search our polygon gazetteer using the tweet's geospatial coordinates to determine which census tract or county or country administrative polygon it falls within. An instance of Solr was loaded with these three data sets using separate Solr cores with a spatial field configured to support very fast point in polygon lookups. The total number of polygons used in our enrichment is about 250,000. The three queries for reverse geocoding enrichment only take about a millisecond each. The Solr field type contains some tuning details of interest:

```
"add-field-type":{
    "name":"geom",
    "class":
"solr.RptWithGeometrySpatialField",
    "spatialContextFactory":"JTS",
    "precisionModel":"floating_single",
    "validationRule":"repairBuffer0",
    "autoIndex":"true",
    "geo":true,
    "distanceUnits":"kilometers",
    "maxDistErr":0.005
}
```

There is also a geometry cache configured in solrconfig.xml as described in Solr's spatial documentation, which we configured to be of size 1024. Another performance factor is our use of EmbeddedSolrServer to embed Solr in the process.

2.3. Apache Kafka

Apache Kafka is a scalable message queue platform designed to handle high data volumes. In this project, the Kafka Streams API is used for consuming from a stream then passing the content to another stream and also for moving data from Kafka into other systems like Solr. Kafka does not support backpressure mechanisms, which creates challenges in some cases like using Kafka to re-index everything. A non- obvious use of Kafka we implemented is that of a persistent, long-term data store or archive. The archival tweets are stored in monthly partitions in a long-term Kafka topic. This approach for archiving has advantages in simplicity but also some serious limitations such as the fact data cannot be sorted or de-duplicated within Kafka as it could in a database. For the purpose of this project we set up a Kafka cluster consisting of 3 zookeepers and 3 brokers. There are 2 consumers for enrichment, an ingester, and 2 topics for input and output. The input topic contains real-time tweets and is fed by the harvester. The tweets from the input topic are consumed by the enrichment and fed to the output topic. Finally, the enriched tweets are consumed from the output topic by the BOP ingester, which transforms the data to a Solr document and then sends it to Solr. Yahoo's Kafka Manager API is used to manage Kafka.

2.4. Solr and Time Sharding

The BOP requirements are a perfect match for date/time-based sharding instead of the default hash based sharding. Sharding refers to the architectural organization of documents into logical slices of the entire dataset such that they can be accessed in parallel. Since Solr has no built-in date sharding mechanisms, a Solr plugin was developed which may be included within Solr in the future. The date sharding plugin is here: https://github.com/cga-harvard/hhypermap-bop/tree/master/bop-core/solr-plugins.

There are two components to the plugin. The primary one is DateShardingURPFactory which examines each tweet as it comes in, and then explicitly sets the special _route_ field based on the tweet's timestamp so that Solr's Distributed URP knows where to send it. The URP will create and remove Solr shards as needed while a tweet is being processed. The other component is DateShardRoutingSearchHandler, which is used instead of the conventional SearchHandler. This piece accepts additional search parameters sent by the client to establish the date range of the search query. The shards parameter is then generated based on applicable dated shards.

2.5. Enhancements to core Solr and Lucene

Apache Solr is an enterprise search server based on Apache Lucene. The following enhancements were made to Solr for the purpose of this project:

2.5.1. LatLonPointSpatialField

LatLonPointSpatialField (LLPSF) is a Solr FieldType that uses both LatLonPoint and LatLonDocValuesField – spatial components in Lucene. Together they are the state of the art spatial fields in Lucene and have shown to be the most performant for searching and sorting point data. LLPSF was released in Solr 6.5: https://issues.apache.org/jira/browse/SOLR- 10039

2.5.2. HeatmapSpatialField

HeatmapSpatialField (HSF) is a Solr FieldType that is expressly designed for heatmaps, unlike the existing RPT type in which it was an after-thought. The core difference between HSF and RPT is that HSF organizes the grid cells by resolution level together (breadth-first tree order), whereas RPT uses a depth-first tree order. High-resolution heatmaps will visit many same-resolution cells and so the intention with HSF is to increase data locality and use simpler cell iteration instead of seeking the Lucene term dictionary to each one.

After deploying HeatmapSpatialField and doing some basic performance measurements, the payoff over RPT was typically negligible in common queries, and there was only a 20% speedup for atypical queries that were already very fast. In essence, the high document count 10M+ per index was such that most time was spent iterating the postings lists (document ID lists) per term (cell), and not so much seeking to each term, despite there being many cells to traverse. Consequently, we didn't feel it was worth contributing this back upstream to Solr. The code for HSF can be found here: https://github.com/dsmiley/lucene- solr/tree/heatmap_hcga.

There are some other differences between HSF and RPT. One is a square grid cell, which is more ideal for data to be displayed than rectangles. Granted this aspect becomes increasingly negligible at higher resolutions. Another is the ability to omit certain resolution levels from being materialized into the index, to save space if you know you won't need them. For example the first level of practical use is not a 4x4 grid but more likely 32x32. These changes may find their way back to Lucene/Solr but have not yet.

2.6. BOP Web-Service

A RESTful API called the BOP Webservice was developed to support our BOP client and potentially other clients. The web service supports keyword search, space and time-based faceting, and CSV export. Having our own API instead of using Solr directly (which would also be possible) makes it much easier for developers to use, and it adds security. The BOP API is based on leading API frameworks Swagger and Dropwizard, and is written in Kotlin, a Java VM based language. The web service supports query/constraints on the following q.* parameters that limit the matching documents:

- q.time: Constraints docs by time range. Either side can be '*' to signify open-ended. Otherwise it must be in either format as given in the example. UTC time zone is implied.
- q.geo: A rectangular geospatial filter in decimal degrees going from the lower-left to the upper-right. The coordinates are in lat,lon format.

- q.text: Constrains docs by keyword search query.
- q.user: Constraints docs by matching exactly a certain user
- d.docs.limit: Limits how many documents to return.

Documents come back sorted by time descending. The response format is text/csv, comma separated with a header row. Values are enclosed in double-quotes (") if it contains a double-quote, comma, or newline.

2.7. BOP Client

The BOP client UI shown in Figure 4 is a browser-based UI with no server component, which makes HTTP/REST requests to the BOP Webservice. HTTP requests are proxied through a Kontena loadbalancer to provide access to the Kontena "weave" network which is one of the first overlay network technologies made available for Docker (Smiley and Kakkar 2017).

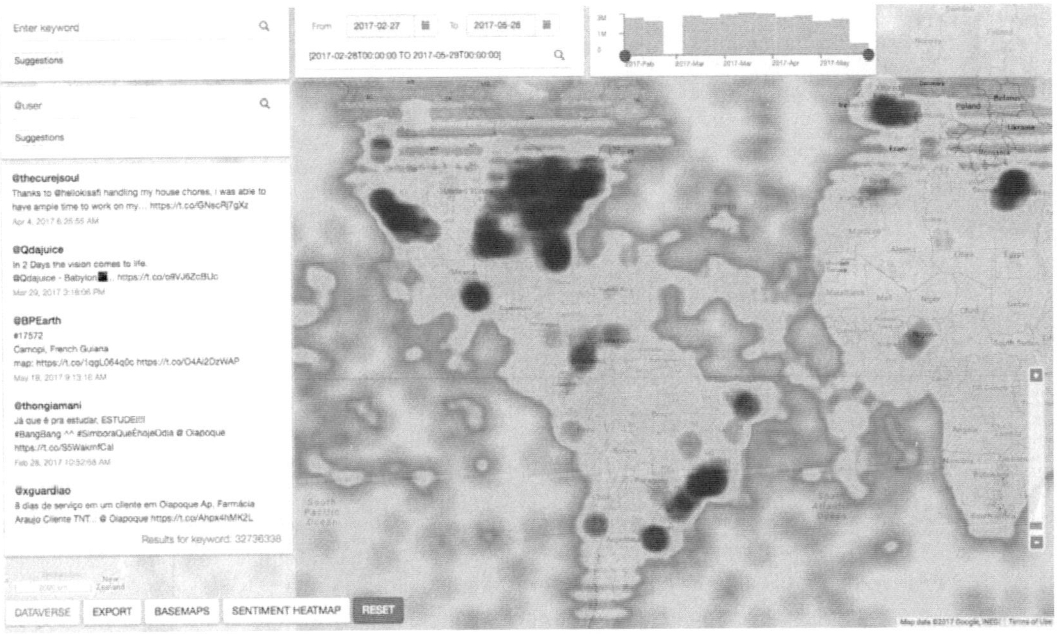

Figure 4: BOP Client UI.

The client UI is built using Angular JS, OpenLayers3, and Node Package Module (NPM). The client UI can adapt to laptop, tablets and phones, and offers the following functionalities:

- Temporal filtering
- Temporal faceting (Histogram)
- Spatial filtering
- Spatial faceting (Heatmaps)
- Text faceting (Tag cloud)

The BOP web service sends out a matrix of data with tweet counts for the whole viewport. OpenLayers 3 with a custom algorithm is used to render that matrix into a heatmap. Rendering uses the Hue Saturation Luminosity (HSL) color model; the different tweet densities affect only the Hue but leaves Saturation and Luminosity unchanged. This helps make sure there are only vibrant colors instead of washed out color when doing the automatic interpolation. In order to be able to spot high-density areas, but also distinguish single tweets in sparse areas we used a sigmoid function to reduce the number of values in the middle classes.

The BOP client UI (UI 2017) allows for downloads of 10,000 tweet IDs at a time along with the enrichment fields in a CSV format. We have developed a tool in Python for rehydration of tweet IDs to fetch full tweet JSON using Twitter REST API's GET statuses/lookup method: https://dev.twitter.com/rest/reference/get/statuses/lookup. The tool returns fully hydrated tweet objects for up to 100 tweets per request, as specified by comma-separated values passed to an ID parameter. This method is useful to get details (hydrate) a collection of tweets. The order of tweets IDS may not match the order of tweets in the returned array. If a requested tweet is unknown or deleted, then that tweet will not be returned in the results list, unless the map parameter is set to true, in which case it will be returned with a value of null. If none of the lookup criteria matches valid tweet IDs an empty array will be returned for map set as false. One must be following a protected user to be able to see their most recent tweets otherwise their status will be removed. Twitter's POST method is preferred for larger requests.

3. Deployment/Operation

The BOP system is hosted on servers run by Massachusetts Open Cloud (MOC), which is a public cloud, based on the Open Cloud eXchange model and is built using OpenStack software.

The BOP runs on 12 virtual server instances: 5 for Solr, 3 for Kafka, 3 for enrichment, 1 as a proxy/gateway. In total it uses about 217 GB RAM and 3500 GB of disk space. Docker/Kontena metrics total 17 services and 133 Docker container instances.

3.1. Docker/Kontena

All components of the BOP are deployed to a Docker based infrastructure managed by Kontena. Docker is a leading software container platform, which provides ease of use: no installation, simplified configurations using environment variables and a common logging mechanism. It is ideal for deploying continuous integrated servers, it is still in its early days and the project requires some risk tolerance to use it in production, which was possible in our case with BOP being a prototype. Kontena was used to deploy Docker, and provides common logging, machine/process statistics, and security. Security is provided at the network or proxy level and is not service specific. Furthermore, Kontena VPN enables the creation of secure networks composed of servers in many locations with local access to all.

3.2. Admin Tools

We used the Yahoo Kafka Manager and the Solr Admin UI to manage Kafka and Solr respectively. The Kafka Manager provides support to manage clusters, topic, partitions and replicas. It also allows one to optionally enable Java Management Extension (JMX) polling for broker and topic level metrics and to filter out consumers. The Solr Admin UI is a web interface that makes it easy to view Solr configuration details, run queries and analyze document fields in order to fine-tune a Solr configuration.

4. Conclusions

Most mapping systems have difficulty representing millions of features on a map unless the visualization has been previously generated. This is because the system either attempts to render an image of millions of features to send to the browser, or it tries to send the features themselves to the browser as vectors. Both approaches are problematic. Rendering images that display millions of features is slow even on a fast server and this work is hard to distribute. Sending large numbers of features to the browser is also not ineffective.

Our approach takes a different tack, one, which attempts to optimize the characterization of the results. The solution is developed around the faceting capability of Lucene/Solr, which is traditionally used for topic counts. This project adapted facet in Solr/Lucene to support geographic counts (Faceting 2017). The approach enables a BOP server to return a spatially referenced array of result counts on the fly very quickly, which the browser then can render as a spatial surface.

The performance of this approach is quite impressive. Heatmap arrays based on queries against hundreds of millions of features are returned within sub-seconds. Speed does vary based on the complexity of the query. We made a mistake in sizing Solr shards by equal time intervals that resulted in large shards before 2016 when geo-tweet volumes were higher. The fix for this will be capping shards by document count as well as time. For earlier years performance is several seconds slower but still impressive when compared to most other mapping systems.

In addition to fast spatial faceting, the BOP also facets on time that enables fast rendering of temporal histograms depicting distribution in time as well as space. In addition to fast results rendering this project developed sentiment analysis (20 ms for tweet with no emoticon and 5 ms for tweet with emoticon) and reverse geocoding (sub-millisecond per poly). It turns out that Solr/Lucene, which is known for blazing text search, is blazing fast for spatial operations as well.

4.1. Future Work

The BOP is a prototype platform. Subject to funding, there are a number of areas, which can be developed, including but are not limited to the following:

- Return arrays of numeric (sensor variables for example) in addition to count values
- A UI to take advantage of census reverse geocoding and faceting to enable fast, on-the-fly spatial analytics using census variable on big data
- Integration with a Spark-based spatial analysis platform such as GeoMesa.

References

API, B., 2017. Bop api. Online, online; accessed May 31, 2017.
URL http://bop.worldmap.harvard.edu/bopws/swagger

Faceting, D., 2017. 2d faceting. Online, online; accessed May 31, 2017.
URL https://cwiki.apache.org/confluence/display/solr/Spatial+Search

Lewis, B., 2016a. Building an open source, real-time, billion object spatio-temporal search platform. Massachusetts Open Cloud Workshop, Boston, MA, USA.

Lewis, B., 2016b. Hhypermap registry: A platform to enable geospatial search. FOSS4G, Raleigh, NC.

OpenStack, 2017. Openstack. Online, online; accessed May 31, 2017.
URL https://www.openstack.org/

Smiley, D., 2016. H-hypermap: Heatmap analytics at scale. Lucene/Solr Revolution, Boston, MA, USA.

Smiley, D., Kakkar, D., 2017. Billion object platform (bop)- an open source, real-time, billion object spatio-temporal search platform. Harvard ABCD-GIS, Boston, MA, USA.

UI, C., 2017. Client ui. Online, online; accessed May 31, 2017.
URL `http://bop.worldmap.harvard.edu/bop/`

Towards a Web-Enabled Geo-Sample Web: An Open Source Resource Registration and Management System for Connecting Geo-Samples to the Web

Anusuriya Devaraju[a,*], Jens Klump[a], Victor Tey[a], Simon Cox[b], Ryan Fraser[a]

[a]*CSIRO Mineral Resources, PO Box 1130, Bentley, 6102 Western Australia, Australia.*

[b]*CSIRO Land and Water, Private Bag 10, Clayton South, 3169 Victoria, Australia.*

Abstract: Within the earth sciences the curation and sharing of geo-samples is crucial to supporting reproducible research, in addition to extending the use of the samples in new research, and saving costs by avoiding sample loss and duplicating sampling activities. In the Commonwealth Scientific and Industrial Research Organisation (CSIRO), researchers gather various geo-samples as part of their field studies and collaborative projects. The diversity of the samples and their unsystematic management led ambiguous sample numbers, incomplete sample descriptions, and difficulties in finding the samples and their related data. These problems are also found in universities, research institutes and government agencies, which usually curate and manage diverse samples. To address this problem, we developed an open source registration and management system to identify geo-samples unambiguously and to manage their metadata and data systematically. The system supports the linking of samples and sample collections to the real world features from where they were collected, as well as to their data and reports on the Web. This paper describes the implementation of the system including its underlying design considerations, and its applications. The system was built upon the International Geo Sample Number persistent identifier system with Semantic Web technologies. It has been implemented and tested with individual users and three sample repositories in the organization.

*Corresponding author
Email address: `anusuriya.devaraju` (Anusuriya Devaraju)

Submitted to FOSS4G 2017 Conference Proceedings, Boston, USA. *September 20, 2017*

1. Introduction

Geo-samples are physical specimens such as rock, soil, sediment, water and vegetation. They are valuable research assets. They are essential for understanding the natural world, required to reproduce scientific experiments, and serve as physical proofs for prior work. For example, groundwater specimens are used to estimate nitrate concentration in groundwater, and biological specimens verify the existence of a species in a particular location. Currently, CSIRO collects and curates a huge number of specimens from various disciplines, such as mineral resources, land and water, energy, agriculture, and manufacturing. These physical specimens also include specimens contributed by other institutions and agencies. For example, the organization manages the National Soil Archive, which is a long-term storage facility of soil specimens from state agencies and national research programs (Karssies et al., 2011). In the future, thousands of samples will be collected to support research activities in the organization.

Geo-samples and their associated data are not commonly available to users other than those that collected the samples. Users may not easily discover them due to several reasons (Lehnert et al., 2006, Devaraju et al., 2016). Sample collectors may follow inconsistent cataloging practices, such as assigning different names or identifiers to the same specimen or the same name or identifier to two or more specimens. They may also make transcription errors while labeling the samples, and the sample metadata may also be incomplete or not recorded in a standard manner. Finally, an online catalog for discovering samples from various sample repositories is not available. To address these issues, we developed an open source resource registration and management system. This development is vital to facilitating the processes of identifying samples, as well as managing and disseminating their information on the Web. The main contributions of this work are as follows:

1. We developed an open source system that supports the effective management and discovery of physical resources such as geo-samples and sample collections[1]. In this paper, we refer to geo-samples and sample collections as a physical resources.
2. We adopted the International Geo Sample Number (IGSN) to support the globally unique and persistent identification of physical resources.
3. We enhanced existing specimen-related controlled vocabularies and developed additional vocabularies. We incorporated the vocabularies into the developed metadata schema to ensure consistency in metadata entries, and on the web portal to improve sample discovery.
4. We demonstrated the application and the relevant of the system in the context of three sample repositories and individual sample curators.

2. International Geo Sample Number (IGSN)

The International Geo Sample Number (IGSN)[2] is a persistent and unique alphanumeric code for identifying physical samples and sample collections. Figure 1 illustrates the IGSN's hierarchical governance structure, which consists of the Implementation Organization of the IGSN (IGSN e.V.), allocating agents, and clients. The Implementation Organization governs

[1] A collection may be a group of arbitrary specimens or large collections of fragments from the same specimen, e.g., a storage tray of drill core sections and rock chips.

[2] `http://www.igsn.org/`

and coordinates standards for identifying and citing physical samples, and operates the international IGSN registration service. The registration service uses the Handle.net System, which is a global persistent identifier resolver service (CNRI, 2010). The Handle.net resolves an IGSN of a sample (e.g., CSCAP876-MJ21)[3] to a landing page[4]. A landing page is the web page that contains the sample descriptions. Allocating agents are the member institutions that are authorized by the IGSN e.V. to register IGSNs within a permitted namespace. For example, CSIRO is one of three IGSN allocating agents in Australia besides Geoscience Australia (GA) and Curtin University. The IGSN e.V. allocated the namespaces 'CS', 'AU' and 'CU' to the allocating agents CSIRO, GA and Curtin University, respectively. Clients (e.g., individual researchers and laboratories) may register their samples with IGSN through an allocating agent. They also maintain the online landing pages of the samples they registered. In the CSIRO implementation, a client sends IGSN registrations to our registration service (allocating agent service) based on the *description schema* we developed. Then, the service forwards the registrations to the international IGSN registration service based on the *registration schema* developed by the IGSN e.V. (Figure 1). The registration schema covers minimal registration information (e.g., sample number, registrant, and datetime), whereas the description schema represents sample information such as identification, collection, curation, and related resources. This separation between registration information and description is important as it gives allocating agents greater flexibility in describing samples for different applications.

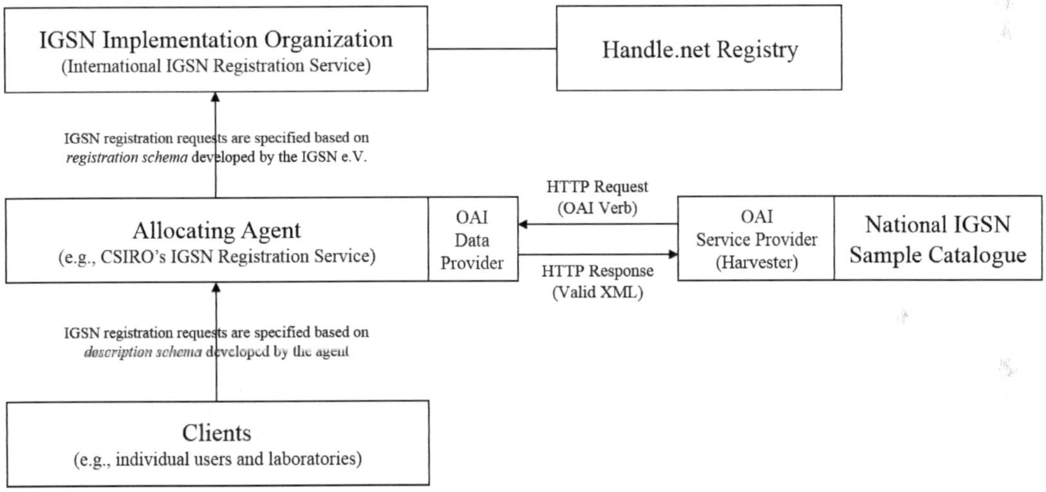

Figure 1: Hierarchical architecture of the IGSN.

3. Resource Registration and Management System

Figure 2 illustrates the architecture of the resource registration and management system. There are two ways to interact with the system. Individual users (e.g., sample collectors) may use a web-based user interface (Figure 3) to register the samples and manage their metadata, whereas sample data repositories may perform the same operations programmatically, e.g., Capricorn Distal Footprints, Repository of the Australian Resources Research Centre (ARRC), and Reflectance Spectra Reference Libraries. Table 1 provides an overview of the samples registered through our system.

[3]To align with practices in DOI, ORCID, etc., IGSN now uses http://igsn.org/{igsn}, redirected to http://hdl.handle.net/10273/{igsn} as the URI from.

[4]https://capdf.csiro.au/data/?igsn=CSCAP876%2dMJ21

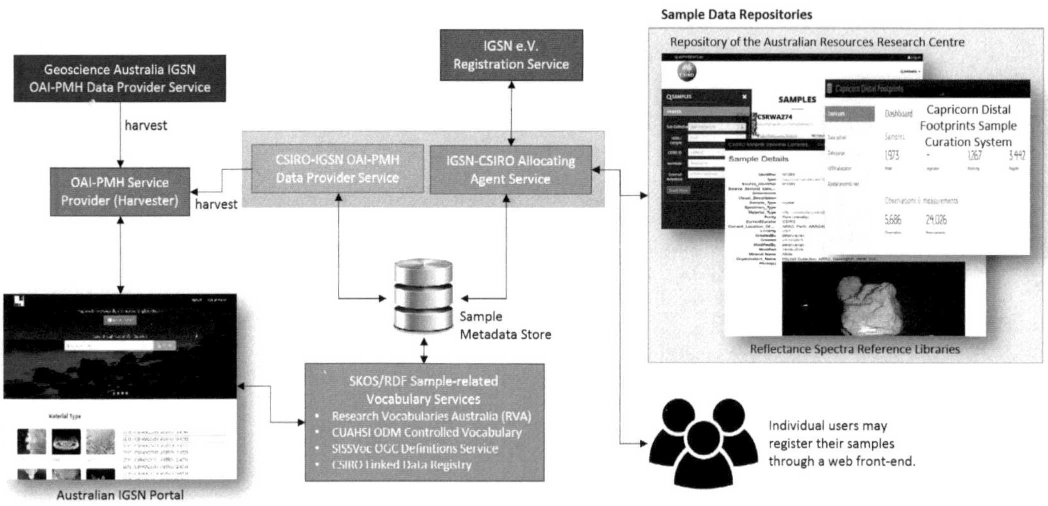

Figure 2: The architecture of the resource registration and management system.

The following are the components of the system developed. The links of the components are listed in Table 3. Their source repositories are available on the Github.[5]

1. **IGSN-CSIRO Allocating Agent Service.** The allocating agent service is a RESTful web service. Clients may authenticate to the service through the HTTP Basic Authentication. The allocating agent service enables clients to register their samples with IGSNs, retrieve and update metadata associated with registered samples, and obtain a list of all the prefixes of the IGSNs registered. Figure 4 illustrates the workflow of a client's sample registration through the allocating agent service.

   In the CSIRO implementation, the IGSN of a sample (e.g., CSCAP876-MJ21) consists of a *prefix* and a *suffix*. A *prefix* is formed by an allocating agent's *namespace*, followed by a *sub-namespace* representing the agent's client. For example, in the prefix 'CSCAP', the namespace 'CS' refers to the allocating agent (CSIRO), and its sub-namespace 'CAP' denotes its client (Capricorn Distal Footprints). The suffix (e.g., 876-MJ21) in the example above refers to the local sample code specified by the client. In our system, clients may only register IGSNs with the sub-namespaces allocated to them. With the hierarchical namespace delegation pattern, we can systematically manage the allocation of specific namespace for different clients in the organization, while at the same time ensuring the global uniqueness of the sample identifiers.

2. **Description Metadata Schema.** The IGSN requests sent by a client to the agent's registration service must be specified in XML, conforming to the description metadata schema we developed (Figure 5). The key features of the metadata schema are as follows:

   - It covers the common concepts associated with geo-samples such as identification, sampling activity, curation, and related resources, so that it can be used to catalogue different sample types in the organization.
   - It supports a batch registration as our applications require large batches of IGSN registrations.

[5]`https://github.com/AuScope/igsn30`

- It has minimal restrictions on mandatory metadata elements (see Figure 5, rectangles with solid border). Some of these elements are required to obtain IGSNs from the international registration service (e.g., *resourceIdentifier* an *landingPage*), while the others are relevant for discovering samples (e.g., *materialTypes* and *curationDetails*).
- It captures the provenance of sample curation as geo-samples are often relocated from repository to another.
- It is also flexible in terms of representing spatial (i.e., coordinates and toponyms) and temporal information (i.e., instants and intervals based on the W3C Date and Time Formats). A coordinate reference system may be specified based on the EPSG Geodetic Parameter Dataset[6].
- It represents several relation types that allow a client to associate a sample or a collection with its related resources through their URIs, such as sub-sample, reference sample, dataset and publication. The information about related resources may exist in the client's sample data curation systems, or in institutional digital repositories such as the CSIRO's Data Access Portal[7] and the Publication Repository[8].

3. **Metadata Store.** The allocating agent service registers IGSNs with the international IGSN registration service and stores the sample metadata (extracted from the description XMLs sent by a client) in a PostgreSQL database. The database also captures client information, such as sub-namespace and authentication details.
4. **SKOS Controlled Vocabularies.** The description metadata schema leverages existing and new controlled vocabularies, which are expressed using the Simple Knowledge Organization System (SKOS). For a list of the vocabularies used by the system, see Table 2. Through the collaboration with the Geoscience Australia, we use the Australian National Data Service (ANDS) Vocabulary Service[9] to develop and maintain the new controlled vocabularies. Concepts in the vocabularies are identified with their persistent URIs in order to ensure machine actionability to the concepts. We include the URIs of the concepts in the description metadata schema in order to provide standardized information about permitted values for specific metadata elements, in addition to ensuring consistency in metadata entries. We also apply the vocabularies for searching and browsing samples on the National IGSN web portal.
5. **Metadata Harvesting and Discovery.** The OAI-PMH is a protocol for harvesting metadata catalogues from digital repositories (Lagoze et al., 2002). In this protocol, a *data provider* offers a catalogue of the repository holdings following the specifications of OAI-PMH, whereas a *service provider* operates a harvester that gathers metadata from one or more digital repositories (Lagoze et al., 2002). We implemented an OAI-PMH data provider service to disseminate the sample metadata records from the metadata store. We also developed an OAI-PMH service provider based on the PANGAEA Framework for Metadata Portals (panFMP) (Schindler and Diepenbroek, 2008). The service provider harvests metadata records from our metadata store through the data provider service, and from other allocating agents, e.g., Geoscience Australia. We developed the National IGSN web portal[10], which provides a common access to the sample metadata records harvested from different allocating agents.

[6] `https://epsg.io`
[7] `https://data.csiro.au/`
[8] `https://publications.csiro.au`
[9] `http://www.ands.org.au/online-services/research-vocabularies-australia`
[10] `http://igsn.org.au`

Table 1: Local sample data repositories and their IGSN registrations (as at 31.05.2017)

Sample Data Repositories	Material Types	Registered samples
Repository of the ARRC	rock, mineral, soil	25652
Capricorn Distal Footprints	rock, vegetation, water, regolith	4232
Reflectance Spectra Reference Libraries	mineral, rock, synthetic material	94

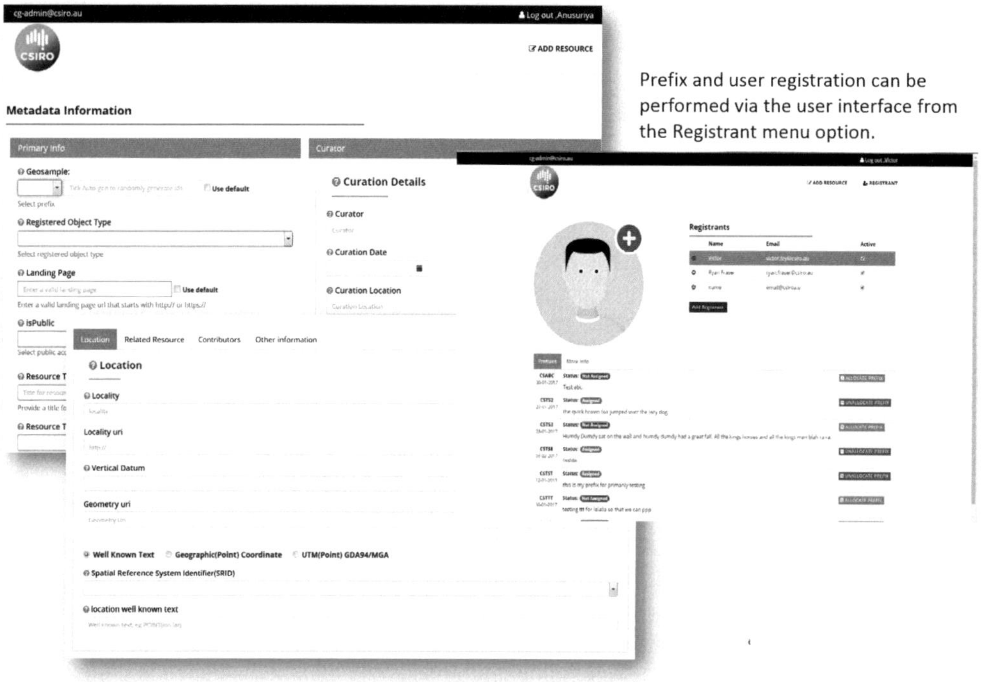

Figure 3: A web-based user interface for registering and managing physical resources.

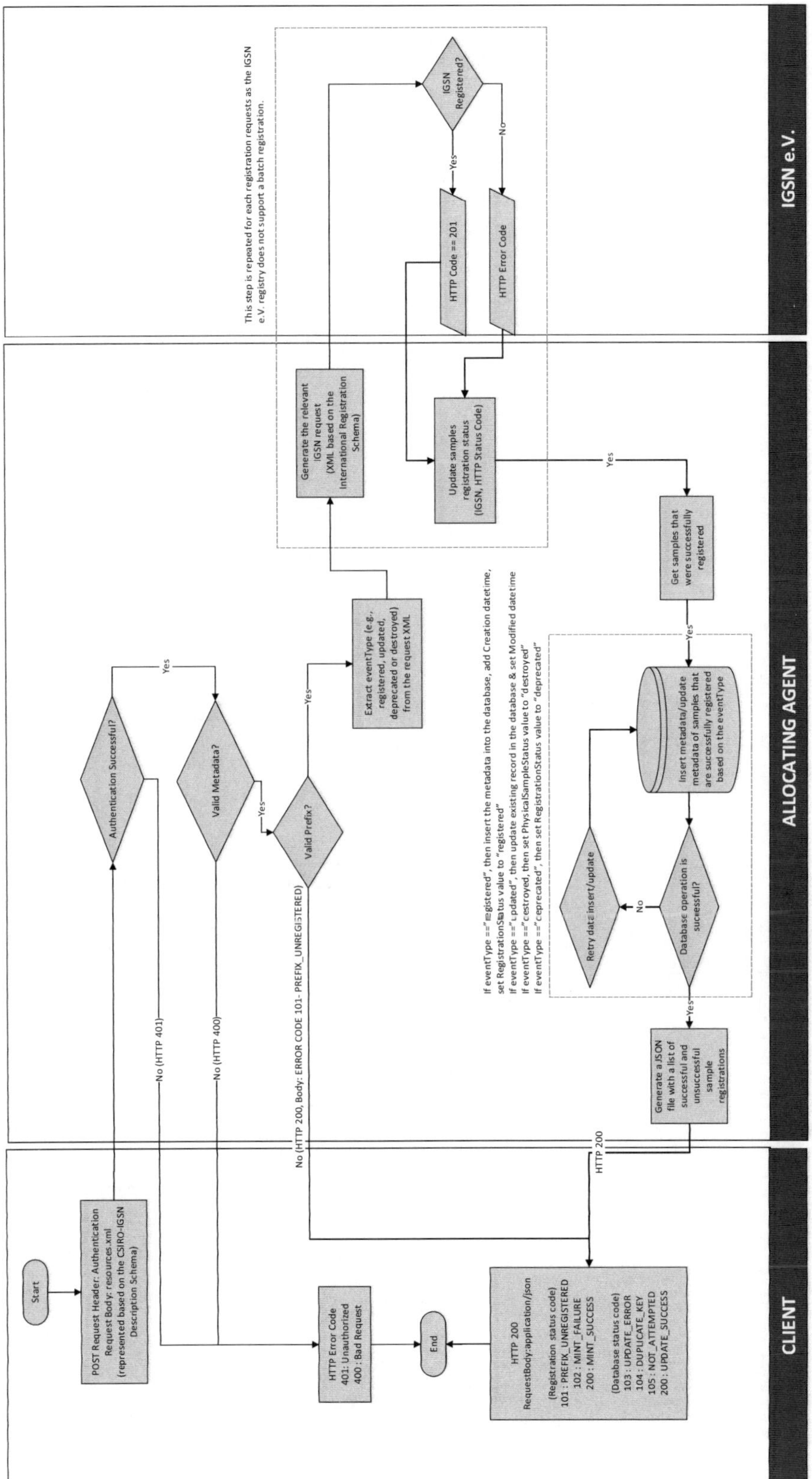

Figure 4: IGSN sample registration.

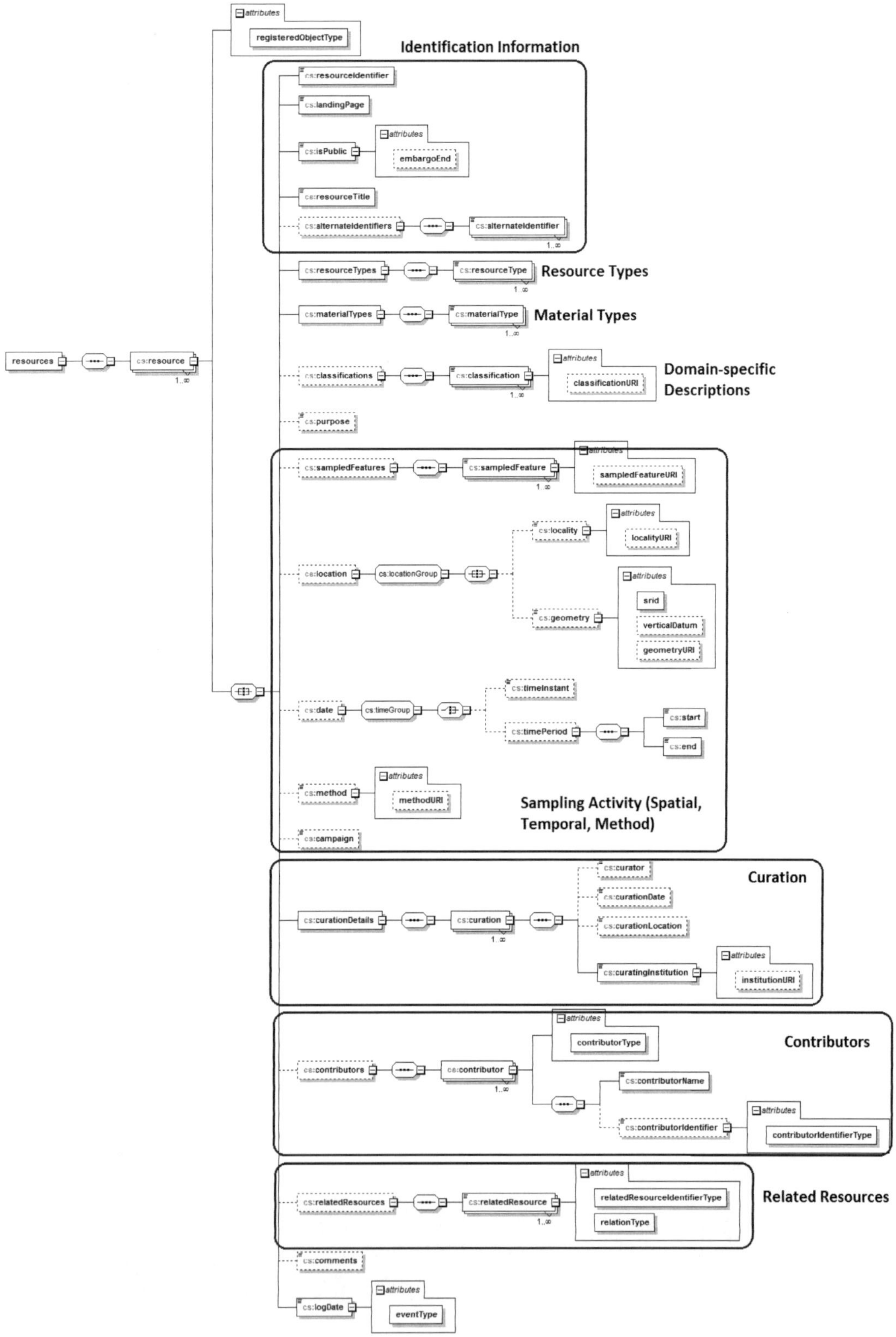

Figure 5: The CSIRO-IGSN description metadata schema (version 3.0).

Table 2: A list of existing and newly developed SKOS vocabularies. New vocabularies are indicated by asterisks*.

Vocabularies	Examples	Service	Provider
Material types[11]	soil, rock, vegetation.	Master Controlled Vocabulary Registry for ODM2	CUAHSI
Specimen types[12]	thin section, grab, dredge, cuttings.	Master Controlled Vocabulary Registry for ODM2	CUAHSI
Nil-reason types[13]	missing, unknown, withheld.	OGC definitions of nil reasons	CSIRO
Contributor types[14]	originator, custodian, point of contact.	Linked Data Registry	CSIRO
*Registration types[15]	physical sample, sample collection, sampling features.	ANDS Research Vocabularies Australia	CSIRO&GA
*Identifier types[16]	DOI, IGSN, LSID, ORCID	ANDS Research Vocabularies Australia.	CSIRO&GA
*Relation types[17]	isDerivedFrom, hasDocument, hasDigitalRepresentation.	ANDS Research Vocabularies Australia	CSIRO&GA

Table 3: A list of the system's components and their links.

Technical Components	Link
CSIRO Allocating Agent Service	https://igsn.csiro.au/igsn30/api
Description Metadata Schema	https://igsn.csiro.au/schemas/3.0/
Metadata Store (source)	https://github.com/AuScope/igsn30/tree/master/sql
CSIRO-IGSN OAI-PMH Data Provider	https://igsn.csiro.au/igsn30/api/service/30/oai
National IGSN Web Portal	http://igsn.org.au
OAI-PMH Harvester and National IGSN Web Portal (source repositories)	https://github.com/AuScope/NatPortalIGSN
Sample Registration and Management Web User Interface	https://igsn.csiro.au/igsn30

[11] http://vocabulary.odm2.org/medium/
[12] http://vocabulary.odm2.org/specimentype/
[13] http://www.opengis.net/def/nil/OGC/0/
[14] http://registry.it.csiro.au/def/isotc211/CI_RoleCode
[15] http://registry.it.csiro.au/def/isotc211/CI_RoleCode
[16] pid.geoscience.gov.au/def/voc/igsn-codelists/identifierType
[17] pid.geoscience.gov.au/def/voc/igsn-codelists/relationType

4. Applications

The system has been tested with individual researchers and three sample data curation systems. The Capricorn Distal Footprints is a collaboration project between CSIRO, government agency, university and industry, which addresses the issue of exploration through cover by examining the geophysical and geological footprints of ore deposits at multiple scales across the Capricorn Orogen in Western Australia (Pearce et al., 2015). The ARRC Repository is a purpose-built facility managed by the CSIRO Business and Infrastructure Services (CBIS). The facility archives all kinds of legacy samples and collections (primarily rocks and minerals) gathered as part of research and development at the organization since the 1970s. The Reflectance Spectra Reference Libraries are used to assist interpretation of hyper- and multi-spectral remote and proximal sensing datasets (Laukamp et al., 2015). The following are applications of the developed system:

Tracking samples from the field to the repository. The CAPDF project uses a mobile field data collection application (Golodoniuc et al., 2017) as part of its field sampling activities. The local identifiers generated by the mobile application are prepended with a relevant namespace (CSCAP), and are then registered through our allocating agent service. This registration process takes place automatically when the records from the mobile application are uploaded into the CAPDF sample data repository. This gives the users an advantage by ensuring consistent use of globally unique sample identifiers after the collection of samples. As a result, samples can be tracked easily at different stages of their life cycle, e.g., specimen handling and storage, laboratory analysis, and eventual disposal.

Asset management. Before we built the system, users had no way of knowing where to look for samples of interest in the sample repository, e.g., ARRC Repository. Each sample collector used their own system to label their samples and sample collections, and this resulted in duplicate or ambiguous labels. Previously, most of the labels on the sample containers were hand-written and without systematic sample identifiers, which may become impossible to place into context after a number of years. As a solution, existing samples and sample collections in the sample repository are registered with IGSNs, thereby allowing their unique identification. We printed QR code labels with relevant IGSNs and affixed them to the sample containers. As a result, users are now able to identify samples in a container by scanning its QR code, which then directs the users to a web page providing the sample information. Users may also use the sample discovery web portal to search for a sample or a sample collection, as well as its storage location (e.g., rack number) in the sample repository.

Web-based discovery of geo-samples and related resources. Previously, there was no way to locate samples from different sample data repositories, unless users searched these repositories individually. Through the sample discovery web portal, users are now able to discover the samples and their associated metadata from both the internal and external sample data repositories. This common access to samples can help to avoid duplicate field trips, create opportunities to re-use existing samples, and promote the reproducibility of sample-based data.

Outlook. Following the successful implementation of the system in CSIRO, we are now collaborating with the John De Laeter Centre for Isotope Research at Curtin University to adapt the system in the context of their Digital Mineral Library.

5. Conclusions

In Earth and environmental sciences, there has been extensive open source development for curating and publishing digital collections (e.g., datasets and documents). Nevertheless, there remains relatively little work on facilitating a globally unique and persistent access to geo-samples and their related data on the Web. To address this challenge, we developed an open source system to support the publication and management of geo-samples and sample collections in CSIRO. It comprises technical (e.g., client applications, metadata schema, metadata store, harvesting capabilities, web portal) and non-technical components (e.g., identifier governance, sample workflows). We built the system to accommodate the needs of both individual sample curators as well as sample repositories. The system assures the uniqueness of the samples, and connects their metadata and data systematically to the Web. The system also offers support to harvest sample metadata records from different sources, which can be aggregated to create new applications, such as the Australian IGSN sample discovery portal.

Through the applications of the system, we demonstrated that it is well-suited for a large organization, in which individual users, projects, and laboratories may all have different requirements for publishing and managing their specimens and collections. Using the development, we can now identify geo-samples unambiguously and discover them easily, and consequently avoid duplicate sampling activities and promote re-use of the samples for new purposes.

Acknowledgments

The IGSN implementation in CSIRO is part of the Research Data Services (RDS) project funded by the Department of Education as part of their Education Investment Fund (EIF) Super Science Initiative. The Capricorn Distal Footprints was funded by the Science and Industry Endowment Fund as part of The Distal Footprints of Giant Ore Systems: UNCOVER Australia Project (RP04-063).

References

CNRI, Dec. 2010. Technical Manual, Handle.Net Version 8.1. Corporation for National Research Initiatives (CNRI).
URL http://handle.net/tech_manual/HandleTool_UserManual.pdf

Devaraju, A., Klump, J., Cox, S. J., Golodoniuc, P., 2016. Representing and publishing physical sample descriptions. Computers & Geosciences 96, 1 – 10.
URL http://dx.doi.org/10.1016/j.cageo.2016.07.018

Golodoniuc, P., Devaraju, A., Klump, J., 2017. Case study: Curation and publication of physical samples using persistent identifiers. In: Geophysical Research Abstracts. Vol. 19. EGU General Assembly 2017.

Karssies, L., Jacquier, D., Wilson, P., Ringrose-Voase, A., Archive, C. N. S., Program, A. C. L. E., 2011. CSIRO National Soil Archive Manual. CSIRO Land and Water.
URL www.clw.csiro.au/aclep/documents/CSIRO_National_Soil_Archive_Manual.pdf

Lagoze, C., Van de Sompel, H., Nelson, M., Warner, S., 2002. Implementation guidelines for the open archives initiative protocol for metadata harvesting. Tech. rep.
URL http://www.openarchives.org/OAI/2.0/guidelines.htm

Laukamp, C., Lau, I., Mason, P., Warren, P., Huntington, J., Green, A., Whitbourn, L., Wright, W., Connor, P., July 2015. CSIRO Thermal infrared spectral library - Part 1: Evaluation and status report 2015. Tech. rep., CSIRO.
URL https://doi.org/10.4225/08/5852dae88924b

Lehnert, K. A., Vinayagamoorthy, S., Djapic, B., Klump, J., 2006. The Digital Sample: Metadata, Unique Identification, and Links to Data and Publications. EOS, Transactions, American Geophysical Union 87 (52, Fall Meet. Suppl.), Abstract IN53C–07.
URL http://abstractsearch.agu.org/meetings/2006/FM/sections/IN/sessions/IN53C/abstracts/IN53C-07.html

Pearce, M. A., Hough, R., Ley, Y., Spinks, S. C., Thorne, R., White, A. J., Golodoniuc, P., Gray, D., Munday, T., 2015. The capricorn distal footprints project: An orogen-scale approach to mineral systems. Proceedings of the GAC-MAC-CGU-AGU Meeting.

Schindler, U., Diepenbroek, M., 2008. Generic xml-based framework for metadata portals. Computers & Geosciences 34 (12), 1947–1955.
URL https://doi.org/10.1016/j.cageo.2008.02.023

Towards OSGeo Best Practices for Scientific Software Citation: Integration Options for Persistent Identifiers fn OSGeo Project Repositories

Peter Lowe[a,*], Markus Neteler[b], Jan Goebel[c], Marco Tullney[d]

[a] *German Institute for Economic Research Mohrenstraße 58, 10117 Berlin, Germany*

[b] *Mundialis GmbH Co. KG Kolnstraße 99, 53111 Bonn, Germany*

[c] *German Institute for Economic Research Mohrenstraße 58, 10117 Berlin, Germany*

[d] *Technische Informationsbibliothek Welfengarten 1B, 30167 Hannover, Germany*

Abstract: As a contribution to the currently ongoing larger effort to establish Open Science as best practices in academia, this article focuses on the Open Source and Open Access tiers of the Open Science triad and community software projects. The current situation of research software development and the need to recognize it as a significant contribution to science is introduced in relation to Open Science. The adoption of the Open Science paradigms occurs at different speeds and on different levels within the various fields of science and crosscutting software communities. This is paralleled by the emerging of an underlying futuresafe technical infrastructure based on open standards to enable proper recognition for published articles, data, and software. Currently the number of journal publications about research software remains low in comparison to the amount of research code published on various software repositories in the WWW. Because common standards for the citation of software projects (containers) and versions of software are lacking, the FORCE11 group and the CodeMeta project are recommending to establish Persistent Identifiers (PIDs), together with suitable metadata sets to reliably cite research software. This approach is compared to the best practices implemented by the OSGeo Foundation for geospatial community software projects. For GRASS GIS, a OSGeo project and one of the oldest geospatial open source community projects, the external requirements for Digital Object Identifier (DOI)-based software citation are compared with the projects software documentation standards. Based on this status assessment, application scenarios are derived on how OSGeo projects can approach DOI-based software citation, both as a standalone option and also as a means to foster open access journal publications as part of reproducible Open Science.

*Corresponding author

Email address: `ploewe@diw.de` (Peter Lowe)

Submitted to FOSS4G 2017 Conference Proceedings, Boston, USA. *September 20, 2017*

1. Research Code, The Workhorse of Science

Software has become the cross-disciplinary workhorse (Brett et al. 2017). Software developed for research is usually created by scientists themselves, instead of through hired professional software developers. According to a survey among 24 universities presented in (BreBrett et al. 2017), scientific software was used by 92% of the researchers, 67% stating that this is fundamental to their research, and 50% developing their own software. This is a potential bottleneck or opportunity for improvement for science, as the coding skills of the researchers directly affect the quality of the scientific results. Effectively, research is becoming dependent upon advances in software, which consists of a diverse range of types of research tools, including operating systems, applications, models, algorithms, middleware and code libraries (Katz 2017). Scientists should be provided with sufficient motivation to hone their programming skills, publish their research software as part of the scientific process, and care about its long term provision to science infrastructure. This motivation should be provided in a suitable coin of the realm of academia, namely recognition for impact and reuse by scientific credit, i.e. citation. This must in turn be footed in reliable social and technical infrastructures. Otherwise, scientific code is in most cases only advanced to the readiness level of a demonstrator/pilot (works for me), as the scope of research projects does not include a drive for code longevity, including refactoring or post-project maintenance.

Scientists should know to find all relevant software packages for his or her field of work, either as a starting point for their individual work, or as a long term repository for the legacy of their resulting research code. This is a challenge, as software projects and code repositories (e.g: GitHub, Sourceforge or Bitbucket) on the web follow lifecycles: They appear, rise in popularity and eventually fall out of use when the user base moves on to next generation software, starting the cycle once again. A pragmatic solution would be to put reasonable trust in software projects, which have been used over a long period in time, having stable user- and developer-communities to ensure updates of codebase and features, reliable and open communication channels and governance models. Creating a personal survey of such state of the art software projects would result in a significant effort for the individual scientist, yet such surveys for geospatial tools have only been infrequently published in journals, like Steiniger and Hunter 2013. A viable alternative can be provided by self-organized federations of software projects, such as the OSGeo umbrella organization which act as impartial arbiters and provide comparable assessments of software alternatives.

2. Open Science

William Gibsons' observation (https://en.wikiquote.org/wiki/William_Gibson) that the future has already arrived, yet remains unevenly distributed holds true for science, libraries and software projects: While some Open Science related paradigms, both on the social and technical level, are already common practice in some fields of science, they remain to be implemented in others. On the positive side, other fields of science have just started to embrace approaches like addressing research software for Open Science. It is possible to benefit from best practices previously developed in these fields, without having to go to the same development curve. As software from geospatial software projects being adopted as novel research infrastructure by neighboring science disciplines like sociology and economics, these sciences can immediately profit by applying best practices for scientific recognition for research software, which did already evolve in other domains.

If at some point the recommendations of Open Science have succeeded to permeate through all sciences, researchers should be trained by, and expected to adhere to these best practices:

- Discovery of existing research code as potential building blocks to serve their research questions.
- Make informed decisions which code building blocks fits their needs best regarding programming language and architecture
- Be able to trust in the quality of code created by others (individuals and groups), as best practices such as coding standards are adhered to.
- Enable others to trust in and reuse one's own research output by transparency and suitable open publication formats.
- Receive due recognition in a suitable currency (i.e. scientific citation), driving career advances including improving code development skills.

These social best practices would be complemented by a scientific work environment based on integrated technology, standards and best IT practices. These include sharing text, data, and code as openly as possible and linking these different types of published research output openly and semantically. Open practices like these support the implementation of a sustainable and open infrastructure (Bilder et al. 2015). Benefits of these desiderata would be an acceleration of the scientific process and scientific discovery, making science a sensible investment for the benefit of the global society.

3. Publishing Research Software

3.1. Open Access Journals for Geospatial Research Software

Several open access journals already provide platforms to address research software topics including the Journal of Open Research Software (JORS) (http://openresearchsoftware.metajnl.com/) and the Journal of Open Source Software (JOSS) (http://joss.theoj.org/). In comparison to the actual magnitude of research code being produced, only a fraction is being communicated via these journals. As a result, advances in scientific software are not being properly communicated and therefore remain inaccessible to other scientists. The effort to write an article about scientific software is not different than for any other scientific topic. Yet what is still lacking is adequate (from the authors' perspective) compensation through recognition by citation. For the time being, scientists continue to focus their efforts on the generally accepted publication channels (i.e. traditional research papers in high impact journals) which provide the desired kind of gratification. A noteworthy exception is the R project community, which has established in 2009 its own open access journal The R Journal (https://journal.r-project.org/) as a scientific publication. This is a best practice example for other project communities.

Even if the number of journal publications on research code could be significantly increased, this will not solve an underlying issue: A journal publication presents a static freeze frame image of a software object, which is often highly dynamic. As Rother 2015 puts it, a publication does not tell you whether the authors are still developing their program further, whether they have stopped maintaining it, or whether the developers have switched fields altogether. From the perspective of a potential user/contributor the following criteria should be applied to assess the quality of research code: Availability on a reliable repository enabling easy installation and source code access, documented recent application, responsive code authors, proof of fitness for use, including sample data and documentation. This could be provided via novel open and responsive formats for journal publications like executable papers for example based on Jupyter notebook technology (http://jupyter.org/).

3.2. Persistent Identifiers for Geospatial Research Software

There are currently no established standards or best practices how to cite research code within a journal publication, like this article. Several approaches to software citation exist, including citations of publications, user manuals, websites, by name, etc. There is also the need to make references to a software project as a general container (e.g. the GRASS GIS project), but in parallel also to refer to specific versions of the code (e.g. GRASS 7.2. versus GRASS 4.2.1). Metadata for software containers and software versions must not be necessarily identical. So independent means are needed to distinctly cite and reference them.

To mitigate this situation the FORCE11 Software Citation group (https://www.force11.org/group/software-citation-working-group) developed software citation principles from data citation principles and existing community practices. These principles comprise the following paradigms: Recognition of scientific software as a legitimate and citable product of research. Scholarly credit for scientific software should be given by credit and attribution. Unique and machine actionable global identification must be ensured, and also persistent use of identifiers and metadata. Access both to identifiable versions of the software and its metadata must be provided. According to (Sperber 2017) the situation is complicated by persistence issues (publications have PIDs, software is not persistent), inherent dynamics of distributed software development, and the absence of standardized (bibliographic) metadata for code.

While the FORCE11 are being embraced as a constructive advance, unresolved issues remain, including the referencing of software containers and versions. The FORCE11 principles call for unique and machine-actionable global identification by persistent identifiers (PIDs). From the user perspective, PIDs enable permanent references to web-objects, regardless whether its web-location should change over time, which would break conventional references by URL.

PIDs are widely used to enable citation of journal publications, research data and research software. Several kinds of PIDs exist, depending on the intended usage scenario (e.g. URN in the German library domain, handles as PIDs with a potentially limited lifespan). PIDs usually point to a human readable landing page, which provides relevant metadata and the most recent link to actual content. Apart from the underlying resolving mechanisms which enable PIDs to link to permanent landing pages, any PID is primarily a social promise to the PID users that this service will be permanently offered. Of the various kinds of PIDs, only Digital Object Identifiers (DOI) currently satisfy the FORCE11 criteria for reliable and machine-actionable software citation. DOIs for research data and software are provided by the DataCite organization (http://www.datacite.org).

The definition of the required metadata provided on landing pages for software is addressed by the CodeMeta project (https://github.com/codemeta/codemeta), which has developed a set of terms to describe software by metadata. These concepts are also used by established repositories and citation services like DataCite, figshare, GitHub, and Zenodo. The terms defined by CodeMeta are intended as a means to ensure the exchange of software descriptions among stakeholders including software users, repositories, communities, citation services, and other parties. They comprise of three object classes: containers, versions, and agents. The latter can be individuals or organizations. Each class has a distinct set of metadata fields. The CodeMeta set of terms is currently not fully consistent with the Software Citation Principles described previously.

4. OSGeo

The OSGeo Foundation (http://www.osgeo.org) is a community driven international umbrella organization, which currently hosts 22 software projects and numerous community projects. Each project is managed and organized by a community of volunteers. Project results like code, data and documentation are provided under free and open licenses for data, software and documents.

Each OSGeo project had to undergo an assessment process referred to as incubation. It assures that a basic set of best practices for open source geospatial projects is met. The current incubation criteria evolved from long duration community software projects. Their goal is to ensure transparent and meritocratic project governance and work practices, i.e. a minimal set standards including openness, and copyright and license, processes, documentation (https://wiki.osgeo.org/wiki/Project_Graduation_ChecklistDocumentation). Being observed by the OSGeo board, the incubation process is not static and evolves over time according to the lessons learned during previous incubation processes and community needs. This makes it a strategic point to introduce and anchor new best practices, like citation principles, within OSGeo for all new projects.

Since 2007, OSGeo has been publishing the online OSGeo Journal (http://www.osgeo.org/journal). It is currently only referenced by ISSN. For the last years, the proceedings of the annual FOSS4G conferences have been published in this journal. Due to the current lack of DOIs, only standard textual references to the proceeding articles can be made. In parallel, a growing number OSGeo-related audiovisual content has been published on several platforms, including video portals such as YouTube and Vimeo. This also includes conference recordings from the FOSS4G conference series. Since 2014, the German national library for science and technology (TIB: Technische Informationsbibliothek) is actively collecting conference recordings from several OSGeo conference series in Europe and North America to ensure their preservation as part of the scientific record (Lowe et al. 2015). The conference recordings are long term preserved, including enriched metadata derived from audiovisual content analysis and receive DOIs which are registered with DataCite. The conference videos are provided through the TIB video portal (https://av.tib.eu). DOIs for audiovisual content allow the citation of both the full video and also selected scenes, which can be referenced by appending a media fragment identifier (MFI) containing time codes to a DOI.

OSGeo has embraced this DOI-enabled service by including it as a best practice for conference video recording in its conference handbook (http://wiki.osgeo.org/wiki/FOSS4G_Handbook). For the OSGeo organization, this has led to a comprehensive archive of content related to its projects and their applications. On the individual level, OSGeo members and other conference attendees are provided with a means to scientifically cite their work, which can be accessed not just through the TIB Infrastructure, but also web-wide search mechanisms provided by the DataCite organization.

5. Case Study: GRASS GIS

The Geoinformation System (GIS) Geographic Resources Analysis Support System (GRASS GIS) is one of the oldest continuously active open source projects, being continuously developed for last 34 years, starting in 1983. The project rules have been improved over time, making it one of the most advanced OSGeo projects and having influenced the shaping of current OSGeo best practices including project incubation.

GRASS GIS currently consists of over 350 modules, which can be combined as needed (reflecting the UNIX toolbox paradigm for modular shell programming) (Neteler et al. 2012). Each GRASS GIS module is required to have a standardized manual page (Figure 1). The manual pages are partially auto-generated from the source code during compilation into machine executable binary programs, leading to a standardized manual style. Manual pages provide a human readable description and metadata of the individual module's functionality, flags and parameters. Latest changes are indicated by a timestamp; each page includes links pointing towards both the recent source code but also to the development history log covering all past versions and authors. If a manual page becomes outdated, additional references to the latest version are included to ensure convenient navigation of software versions.

New functionality for GRASS GIS is being added continuously, often as the result of research projects, or as part of campaigns such as code sprints or Google Summer of Code. This is often done by early career scientists. The transfer of such add-on code from the private domain of the scientist into the project software repository, and from there possibly onward into other such OSGeo projects is done in defined steps. From the perspective of the project, the initial step is the code deposition in a repository for add-on modules, which requires the granting of write access following a public request by the coder. The coder agrees to adhere to minimal standards (https://trac.osgeo.org/grass/wiki/Submitting) for code and documentation. Add-ons are not part of the canonical GRASS distribution, but can be downloaded and automatically installed within a GRASS release version. In a second step, after a review of the add-on's code structure and confirmation of general interest in its functionality, the add-on module can enter the core codebase, becoming a canonical functionality in GRASS GIS.

Once a sufficient amount of development goals have been reached at a given point of time, the ongoing GRASS development activities on a particular version are temporarily stopped ("frozen") by the developer community. Following a testing phase, the code snapshot of the actual project codebase is released as a new production grade GRASS GIS version, assigned with a distinctive version number and provided via the GRASS GIS Website. All versions of the GRASS code base since 1990 are provided in the distributed OSGeo server infrastructure which ensures reliable access and documentation to the evolving versions of the code base, allowing for retrospective transparent access, in the sense of Open Science. While users can also directly access the latest version of the unstable development code base in the repository instead of the officially published releases, this practice is discouraged due to the current fast pace of production grade releases (Neteler 2014). Despite GRASS GIS add-ons can be published anywhere in the internet, the authors are encouraged to make use the of official GRASS GIS add-ons repository to become part of automated provision of installable executables, to be included to the list of add-ons on the main project website and to become part of the source code peer-review mechanisms established within the project.

Currently, the GRASS GIS project homepage provides advice, how to cite versions of GRASS GIS in scientific publications and also provides a BiBTeX entry (https://grasswiki.osgeo.org/wiki/GRASS_Citation_Repository). In parallel, GRASS related code publications have started to appear on the open access research data repositories like Zenodo (https://zenodo.org/search?page=1size=20q=GRASS%20GIS) since 2015, which provide citation via DOIs.

From a technical perspective, the GRASS GIS codebase can be hosted on a variety of repository platforms: While copies of the GRASS codebase are available via the Git repository hosting service GitHub (https://travis-ci.org/GRASS-GIS/grass-ci), the primary GRASS code repository uses Apache Subversion (https://subversion.apache.org/) as its central software versioning and revision control system, hosted within the OSGeo IT infrastructure. Currently

GRASS add-on modules published on external repositories are not automatically merged with the primary GRASS repository, so sharing of such external-hosted GRASS code would require personal effort by the developers by submitting it to the main GRASS repository to benefit from the project's code review and testing.

A discussion of the usage of PIDs for software citation picked up in 2016 within the GRASS Project Steering Committee
(https://lists.osgeo.org/pipermail/grass-psc/2016-November/001606.html,
https://lists.osgeo.org/pipermail/grass-psc/2017-January/001628.html). Recent software repositories used in the social sciences were proposed as analogies and potential best practices for the GRASS GIS community. In addition, an exchange with the DataCite organization introduced the option of either the GRASS GIS project or the OSGeo umbrella organization becoming DOI registration services themselves, enabling novel citation services within the OSGeo communities. According to DataCite, the GRASS project infrastructure, meets and exceeds the minimal requirements for DataCite-approved DOI-registering infrastructures. Manual pages for GRASS modules already meet the requirements for DOI landing pages regarding identifier, creator, title, publisher, publication year and resource type (Figure 2). The discussion addressed the variety of approaches for software citation currently existing. To give the GRASS project's software due recognition (by easing scientific citation), several use cases have to be considered: citing the project as a container, irrespective of a particular software version (i.e. GRASS GIS),

1. citing a specific version (e.g. GRASS GIS 7.2), and
2. citing a particular version of a GRASS module (e.g. the g.access

(https://grass.osgeo.org/grass72/manuals/g.access.html) module in GRASS GIS 7.2.). For this purpose, a consistent approach is needed to cite a GRASS module when it is considered merely an add-on module and also after it has been included in the main GRASS functionality portfolio. The handling of software versioning has already been introduced by the Zenodo repository (http://blog.zenodo.org/2017/05/30/doi-versioning-launched). Consensus within the GRASS project community on how to proceed remains to be reached.

The current GRASS GIS repository meets these requirements for references to all published production versions of GRASS GIS and individual modules. This is demonstrated by the visual analysis (software container level) by (Neteler 2011, Figure 2) showcasing the evolution of the complexity of the GRASS release versions over 12 years, between 1999 and 2011, including the agents (individuals) as suggested by the CodeMeta project.

6. Summary and Outlook

As journal publications continue to be the standard form of scientific exchange, it is crucial to support the change of citation culture towards a more prominent recognition for code and data to foster Open Science. For geospatial research, both the GRASS GIS software project and the OSGeo umbrella organization have the capabilities to adopt DOI-based software citation due to advanced status of their internal best practices. OSGeo has already embraced DOI-citation as best practice for conference recordings since 2016. DOIs for code submissions can become an additional infrastructure offering for developers of research software. This will motivate submissions of additional research code to OSGeo projects and also foster journal publications about the code.

The approach towards DOIs for software within OSGeo needs to be independent from the current software repository situation, as it must be open towards future changes of underlying technical infrastructures. Bottle-necks and potential vendor lock-in must be avoided, as in the past.

As a follow up step, acceptance and usage of unique identifiers also for persons, like ORCID, can become a strong driver to distinguish and trace code submissions by individuals and its reuse as altmetrics, based on open standards and irrespective of underlying technical infrastructures. While such mechanisms are currently being developed for data and code citation by DataCite, they are already provided by the Crossref organization for DOI-based journal publications and ORCID (https://www.crossref.org/blog/crossrefs-doi-event-tracker-pilot/).

Based on the current state of technology and established best practices within OSGeo, the following scenarios appear feasible for the near future:

DOI-application within the GRASS GIS Project and impact on OSGeo (see Figure 3 for overview):

1. The GRASS GIS project has DOIs minted both for the overall GRASS GIS container and also individual code releases (e.g. GRASS7.0, GRASS7.2, etc.). This requires limited effort as only a very small number of DOIs is required.
2. In addition, individual GRASS modules also receive DOIs for each new version. The manual pages of the modules serve as landing pages. For this, recommendations by the Research Data Alliance (RDA) regarding the citation of evolving data could be applied to evolving code (Rauber et al. 2015).
3. Best practices for other OSGeo projects are derived from the lessons learned by the discussion within the GRASS project regarding DOI use.
4. Software citation by PID or other means is included in the OSGeo incubation roster for project quality assessment.

Management of DOI registration for OSGeo projects:

1. The GRASS GIS project becomes accredited as a data center via an allocating member of DataCite. Such a member organization could be like TIB or USGS. Other OSGeo projects can become independently accredited by other DataCite allocating members, to receive their own DOIs. All these DOIs will not be distinguishable as being related to OSGeo projects.
2. OSGeo joins the DataCite organization as an allocating member and becomes able to mint an unlimited amount of DOIs with its own distinctive DOI-prefix . Such OSGeo-DOIs can be provided to all OSGeo community projects for code, data and documentation. An example for this approach within DataCite is CodeOcean (https://search.datacite.org/members/ocean), which serves both as data center and DataCite allocating member.

In addition to the described bottom-up options and needs which have been described for the GRASS GIS project and OSGeo, top-down support by international research organizations and similar bodies is needed, to ensure that Open Science becomes an established paradigm in science: Openness and sustainability in science require an infrastructure of reliable future-safe repositories for data and code, but also the acknowledgement of their significance as scientific output. This significance needs to be reflected in the overall reputation system, to ensure viable incentives for early career scientists, to ensure that their code and data are properly published.

NAME

g.access - Controls access to the current mapset for other users on the system.
If no option given, prints current status.

KEYWORDS

general, map management, permission

SYNOPSIS

g.access
g.access --help
g.access [group=*string*] [other=*string*] [--help] [--verbose] [--quiet] [--ui]

Flags:

--help
Print usage summary
--verbose
Verbose module output
--quiet
Quiet module output
--ui
Force launching GUI dialog

Parameters:

group=*string*
Access for group
Options: *grant, revoke*
other=*string*
Access for others
Options: *grant, revoke*

DESCRIPTION

This program allows the user to control access to the current mapset. Normally, any user can read data from any GRASS mapset. But sometimes it is desirable to prohibit access to certain sensitive data. The *g.access* command allows a user to restrict read and execute access to the current mapset (see UNIX *chmod* command). *g.access* will not modify write access to the current mapset.

The user may, for example, allow only users in the same UNIX group to read data files in the mapset, or restrict the mapset to personal use only.

NOTES

Under GRASS, access to the mapset PERMANENT must be open to all users. This is because GRASS looks for the user's default geographic region definition settings and the location TITLE in files that are stored under the PERMANENT mapset directory. The *g.access* command, therefore, will not allow you to restrict access to the PERMANENT mapset.

The *g.mapsets* command isn't smart enough to tell if access to a specified mapset is restricted, and the user is therefore allowed to include the names of restricted mapsets in his search path. However, the data in a restricted mapset is still protected; any attempts to look for or use data in a restricted mapset will fail. The user will simply not see any data listed for a restricted mapset.

SEE ALSO

UNIX manual entries for *chmod* and *group*
g.mapsets

AUTHOR

Michael Shapiro, U.S. Army Construction Engineering Research Laboratory

Last changed: $Date: 2011-11-08 13:24:20 -0800 (Tue, 08 Nov 2011) $

SOURCE CODE

Available at: g.access source code (history)

Main index | General index | Topics index | Keywords index | Graphical index | Full index

© 2003-2017 GRASS Development Team, **GRASS GIS 7.2.2svn Reference Manual**

Figure 1: Example of a GRASS GIS standardized manual page for a module (g.access module).

This type of manual page can also serve as a landing page for a DOI, including a high level description, metadata, and links to the recent source code and previous versions. The module is an example for code which has been maintained by the project community for a long time after the original developer left the project (GIS 2017).

Figure 2: Still image from GRASS GIS source code evolution visualization (Neteler, M. 2013: https://doi.org/10.5446/14652t=00:52,09:33).

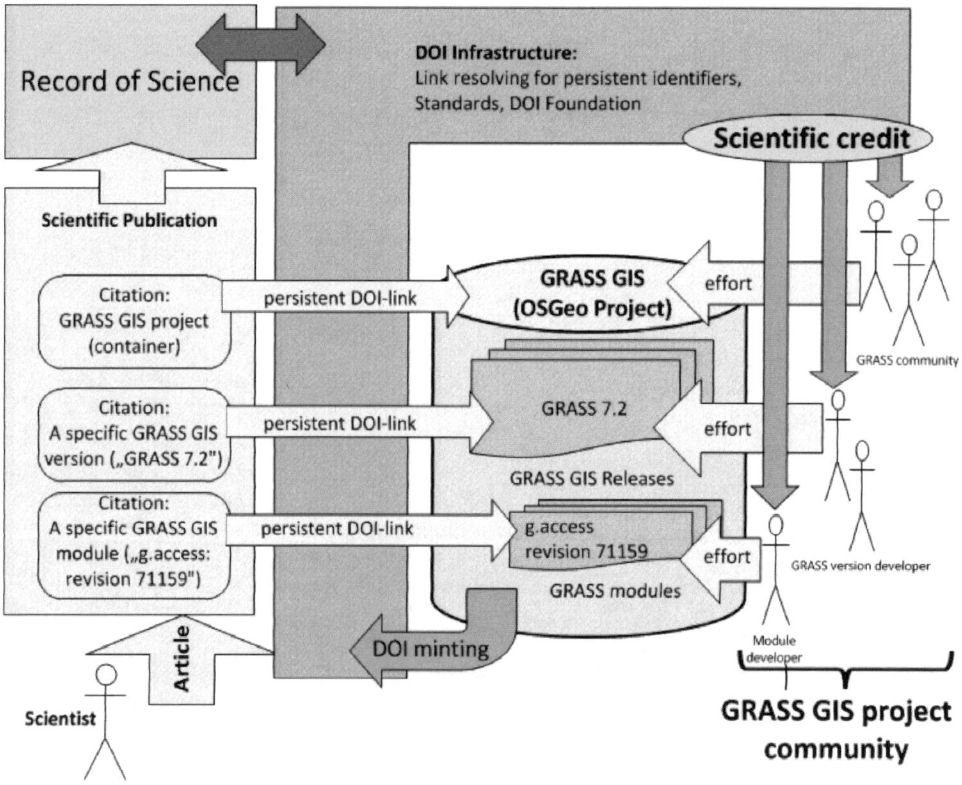

Figure 3: Proposed adoption of DOIs for software by the GRASS GIS project (high level view).

The animated visualization of the source code development (generated with the Gource software, http://gource.io/) shows as a tree the software development over time. In the beginning, the center is the root directory of the initial software project which then evolves over time. Newly created subdirectories are shown as branches with their files as leaves. The activity of software development is shown through small agents in the visualization which represent the various developers working on the source code over time. The GRASS GIS development visualization video summarizes fifteen years of development in which a major source code reorganization happened.

This high level view highlights the requirements, resulting new capabilities and emerging benefits resulting from the adoption of Digital Object Identifiers (DOI) by the GRASS GIS project. Once the GRASS GIS project starts minting DOIs via a member of a DOI registration agency, these DOIs can be used for scientific citation in journal publications and books, which become part of the record of science. The DOI infrastructure enables persistent links leading to the landing pages for software objects, on the container-, version- and code module-level. This allows to cite, and thereby provide due scientific credit, to the individuals who are maintaining and developing the codebase of the GRASS GIS project. The DOI infrastructure also enables the monitoring of citations, i.e. the re-use and possible extension of the cited software as a measure of the impact of the effort of the software authors.

References

Bilder, G., Lin, J., Neylon, C., 2015. Principles for open scholarly infrastructure-v1. Online, online; accessed May 31, 2017.
URL `http://cameronneylon.net`

Brett, A., Croucher, M., Haines, R., Hettrick, S., Hetherington, J., Stillwell, M., Wyatt, C., 2017. Research software engineers. State of the Nation Report, Zenodo.

GIS, G., 2017. Historical notes. Online, online; accessed May 31, 2017.
URL `https://grass.osgeo.org/home/history/`

Katz, D. S., 2017. Software citation: a cornerstone of software-enabled research. 2nd Conference on Non-Textual Information: Software and Services for Science.

Lowe, P., Neumann, J., Plank, M., Ziedorn, F., Lozar, R., Westervelt, J., Inman, R., 2015. Grass gis, star trek and old video tape. OSGeo Journal 14, 43–48.

Neteler, M., 2011. Grass gis 6.4 development visualization from 1999 to 2011 with gource. GRASS Development Team.

Neteler, M., Bowman, M. H., Landa, M., Metz, M., 2012. Grass gis: a multi-purpose open source gis. Environmental Modelling and Software 31, 124–130.

Rauber, A., Asmi, A., van Uytvanck, D., Proll, S., 2015. Data citation of evolving data - recommendations of the working group on data citataion (wgdc). Online, online; accessed May 31, 2017.
URL `https://www.rd-alliance.org/rda-wgdc-recommendations-extended-description-tcdl-draft.html`

Rother, K., 2015. How to recognize good scientific software? Online, online; accessed May 31, 2017.
URL `http://www.academis.eu/posts/good_software`

Sperber, W., 2017. Lost in the web? new approaches for information on mathematical software. Technische Informationsbibliothek (TIB).

Steiniger, S., Hunter, A. J. S., 2013. The 2012 free and open source gis software map - a guide to facilitate research, development and adoption. Computers, Environment and Urban Systems 39, 136–150.

Tracking 19th Century Late Blight from Archival Documents using Text Analytics and Geoparsing

Laura Tateosian[a,*], Rachael Guenter[b], Yi-Peng Yang[a], Jean Ristaino[1]

[a]*Center for Geospatial Analytics, North Carolina State University*

[b]*Department of Entomology and Plant Pathology, North Carolina State University*

Abstract: In the year 1845, Ireland's potato crop was struck by a devastating disease. The potato disease that killed Ireland's crop caused devastation for seven long years and led to mass starvation and emigration from the country. The cause of the potato destruction was a fungus-like plant pathogen later named Phytophthora infestans (DeBary 1876). There are several theories about the origin of the disease and the source of the 19th century outbreaks. We use historical documents contemporary to that time to investigate spatial information that might inform these mysteries. We present methodologies for automatically extracting information from these voluminous data sources. We identify and map geographic locations that are proximate in the text to key terms related to potato blight. Data sources include agricultural documents with extensive discussions of crop yields and failures, seed export and import, and weather conditions, along with location names. We apply natural language processing tools to automate text mining of the data within narrative passages. Specifically, we used text analytics tools from the Natural Language ToolKit (NLTK) and location name extraction from Berico's CLAVIN geoparser. We couple these to mine the relationships between locations and reports of potato disease. Results are displayed in an interactive Web mapping tool for users to spatially explore the pertinent data for trends in the emergence of 19th century late blight.

*Corresponding author
Email address: `lgtateos@ncsu.edu` (Laura Tateosian)

Submitted to FOSS4G 2017 Conference Proceedings, Boston, USA. *September 21, 2017*

1. Introduction

The movement of plant pathogens into new geographic areas is a major factor in the emergence of novel virulent lineages that threaten food security. A tragic example of a food crisis occurred in Ireland 1845 when potatoes, a diet mainstay at that time, were struck by a devastating plant disease that plagued the country for seven years and resulted in mass starvation known as the Irish Potato Famine (Bourke 1964). In fact, initial disease outbreak in potatoes occurred two years earlier in 1843 in the US. The cause of the potato destruction, was later identified as a plant pathogen, named Phytophthora infestans (P. infestans) by DeBary 1876, commonly referred to as 'late blight'. A yet unresolved question is the geographic origin and patterns of spread of the disease for the 19th-century US and European outbreaks and the means by which it spread. For example, it may have come from potato seed tubers imported from Mexico or infested fertilizer from Chile.

Several maps that contain putative information about the spread have been created (Bourke 1964; Yoshida et al. 2013, Birch and Cooke 2013). However, these map provides only a tentative visual of the location of the disease in the years 1844 and 1845 and the direction of spread and they are based on manual interpretation of published reports or tracking of genetic lineages. It is important to develop more sophisticated methods to track emerging plant disease outbreaks, methods that combine both unstructured metadata and genotyping so that systems can be deployed more rapidly to monitor current plant diseases from various data sources. It is known that the disease was first introduced into the US around the ports of New York and Philadelphia. However, the source of the infected potatoes is still widely debated with some suggesting an Andean origin while others suggesting a Mexican source (Gomez-Alpizar et al. 2007, Goss et al. 2014, Martin et al. 2015).

Documents archives containing agricultural reports and news articles from that time can help to answer these questions. The volume of data necessitates data mining techniques to filter the texts for potentially relevant information. The documents are narratives, not easily machine-readable structured tabular data. Hence, natural language processing tools are needed to extract information from these documents. The questions of disease origin and spread are spatial, so GIS plays a central role in this investigation. In this paper, we develop a foundation for answering questions about disease origin and spread based on unstructured documents. Our goal is to identify patterns related to various disease-related parameters, including the geographic place names in the text. We use open source software tools to extract geo-related terms and other useful relationships in historic documents and we generate interactive maps based on the results.

To learn more about the agricultural conditions, we need documents contemporary to the time with a focus on crops. With this in mind, we selected our main datasets. The annual US Patent Office Reports 1841-1850 are specifically targeted for their content discussing agricultural current events and commerce in the US in the 1840s (US Patent Office 1841-1850). In this period (prior to the formation of the US Department of Agriculture), detailed agricultural reports were part of the US Patent Office reports. These documents contain over 6,000 pages of information about the nature, cause, and spread of the disease. The conversations around potato agriculture in the US at that time are of particular interest since the disease appeaedr there two year prior to the European breakout. A secondary dataset, the Google Books Corpus (Jean-Baptiste and et. al. 2011), containing millions of books written over hundreds of years provides some context for our patent corpus with a broader view of literature.

The tools we used for our work reside loosely in four categories, text cleaning and searching, natural language processing, geoparsing, and GIS:

• *Text cleaning and searching.* We used the Python 'PDFMiner' and 're' (regular expression) module and modified Peter Norvig's Python spelling corrector module. PDFminer module can be used to export Portable Document Format (PDF) documents to ASCII text files. The re module can be used to search for substring patterns in text.

• *Natural language processing.* We used the Python Natural Language Toolkit (NLTK) and the Google Ngram Viewer. NLTK is a highly-developed well-documented Python library for working with human language data. Google Ngram Viewer, a powerful online search engine that charts frequencies of words. N-grams are contiguous substrings of length n (E.g., for search phrase, *potato rot*, N=2, and the term potato not followed by *rot*, would not be selected).

• *Geoparsing.* We used CLAVIN and GazeGIS. Geoparsing is the process of identifying and assigning a geographic coordinate to toponym instances in unstructured text (i.e., text, such as our agricultural narratives, that is not structured in any predictable format). Geoparsing uses a gazetteer (a dictionary of place names) and entity extraction to automatically detect place names. Then it uses a set of heuristics to resolve the location of non-unique place names based on the context of the surrounding passage. We used the open source geoparser, CLAVIN. (Greenbacker 2013). GazeGIS (Tateosian et al. 2017), a software program that combines geoparsing and web mapping to extract and map location names from texts, provides a graphical user interface for geoparsing.

• *Mapping.* We used Google Maps, Google Visualization API, and Google Sheets to create interactive maps of the data.

In the upcoming section, we explain how we use these tools to explore our data. Then we present and discuss our results before we conclude with plans for future work.

2. Methodology

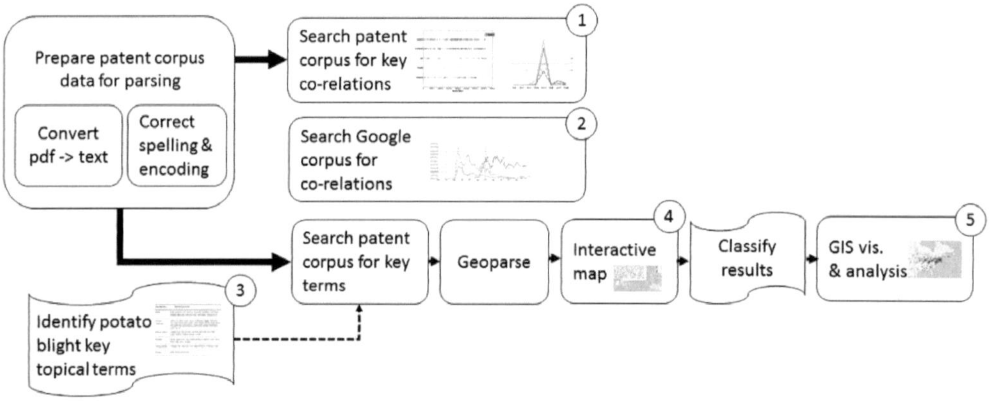

Figure 1: Overview of the methodology.

Fig. 1 provides an overview of our methodology with the deliverables from this process labeled 1-5. First, to perform machine analysis of the patent reports, we need to preprocess the documents to make them suitable for machine processing. Then, deliverables 1 and 2 summarize pertinent characteristics of the patent document dataset and the Google Book corpus respectively. Our third line of investigation is informed by deliverable 3, a list of terms related to the topic of potato disease. This feeds the main thrust of our work, which searches for key terms, and then identifies and maps geographic terms within these search results. The following subsections step through the workflow with details of the methodology.

2.1. Preparing documents for parsing

For our primary dataset, the original US Patent Reports documents from the years 1841-1850 were scanned with optical character recognition and stored as PDF documents.. Since machine processing requires text format, we used the Python PDFminer module to export the documents as ASCII text files.

The conversion of PDF documents to text introduces some spelling and encoding artifacts. First, some characters are encoded as ISO-8859-1 characters which confound the NLTK processing. The second issue arises from working with historical documents. The original documents had sustained some damage due to age before they were digitally archived. The paper had dark marks that appear in the scanned digital format and confound the PDF to ASCII converter, which interprets some stray marks as punctuation, resulting in misspellings (e.g., potato might become pot;ato due to a mark near the word). Third, the original line breaks are not preserved during the conversion, resulting in two types of issues. In cases where the line does not correspond to the end of a sentence, word pairs are run together or a hyphen is embedded within a non-hyphenated word. We built upon Norvig's Python spelling corrector and used the re module to remove the poorly encoded characters and repair spelling issues.

2.2. Summarizing the data

The reports in the corrected text file format were then loaded into NLTK for text analysis and exploration. To explore trends and extract relevant information about potato disease from the patent reports, we used NLTK to view the concordance, collocation, and text dispersion plots for a few core search terms, such as 'potato', 'disease', and 'rot'. A concordance view shows every occurrence of a given word, together with some context. A collocation is a sequence of words that occur together unusually often. A text dispersion plot determines the locations of a word in the text (how many words from the beginning it appears) and plots a glyph in each location on a graph.

We also wrote our own custom concordance keyword searcher (KeyTermSearcher), again using the re module, to extract paragraphs around key terms and tally the counts while omitting some undesired collocated word pairs, such as 'sweet potatoes', which are not of interest to our research. The datasets output from this script is used in the methods described in Section 2.4. A broader search beyond the patent reports of published literature was done using the Google Ngram Viewer to find keywords in the collection of Google Books. Google Ngram can search all books available through Google, dating back to 1800. The program then graphs the frequency of search terms over a period of specified years. To look for trends in the use of potato disease related terms over time, we searched 200 years of documents to plot the usage of the terms 'potato disease', 'potato rot', and 'late blight'.

2.3. Extending the key term searches

The core idea behind our research is to locate discussions related to potatoes and potato diseases and find the place names that they reference. So we needed to identify a set of terms to drive these searches. We are interested in the origin of the pathogen and movement of the disease, how the farmers managed the onset, and contributing environmental factors such as weather conditions. Based on literature review of the disease and through inspection of our preliminary concordance results, we identified six categories with lists of key terms (Table 1).

Potatoes are referenced generically as 'potato'. but are also referred to by cultivar (e.g., chenango, lady finger or blue-nose). Mention of seed or seed sets was also important, as this indicates in some cases, the origin of the seed tubers. We also identified similar but irrelevant terms (such as, *sweet potatoes*) to omit. Potato diseases are variously referred to descriptively by symptoms (e.g., *black spots*) or with synonyms for disease (e.g., *malady, murrain*). The term, *dry rot*, is the name for another disease, so this is omitted.

Classification	Related Keywords
Potato	potato, potatoes, seeds, seed set, Irish potato, Chenango, Lady fingers, Solanum tuberosum, blue-nose (omit: sweet potato, sweet potatoes)
Disease symptoms	mould, rot, rotted, disease, decay, sudden decay, fungus, black spot, loss, blight, black spots, malady, potato cholera, murrain, potato plague, evil, potato evil, potato disorder, potato rotted, mildew, rusted decay (omit: dry rot)
Disease control	sulphuric acid, salt, salt water, sea water, lime-sulfur, lime sulfur, copper sulphate, bordeaux mixture, sea salt
Weather	rainfall, temperature, cold, rainfall amount, precipitation, moist, freeze, flood, damp, storm, drought
Transportation	shipping, ship, cargo, port, river, imported, import, waterways, canal
Storage	cellar, buried, preservation

Table 1: Key Search Terms Classifications (deliverable 3 in Table 1).

Disease control methods, transportation and storage of potatoes, and weather also have a direct relationship with the onset, manifestation, and movement of late blight. Various methods were used to attempt to control the disease (e.g., *sulphuric acid, salt water*). Weather conditions affected the severity and spread of the disease. Potatoes were transported on waterways and seeds were imported from various parts of the world. Storage conditions also play a role.

We anticipate that iterative searching and mapping will generate additional terms of interest. The integration of these many terms will help to build a spatial understanding of where the disease was present and how it was spreading across the globe.

2.4. Geoparsing and Mapping

The core goal of this work is to extract locational information about the origin spread of the potato blight. We extract related information, find the geographic references in that text and then map the results so that users can validate the results and review only the most pertinent to identify promising leads. To implement this, we use key terms from Table 1 as input to the KeyTermSearcher process described in Section 2.2. This extracts the terms and about 2000 characters of surrounding text. To find mentions of place names we load these selections into GazeGIS, which calls CLAVIN geoparser and extracts the place names, their geographic coordinates, and their position in the text.

We then map the locations in an interactive Google Map linked to a Google Sheet of the key term passages. The interactive map (deliverable 4 in Table 1) is designed so that users,

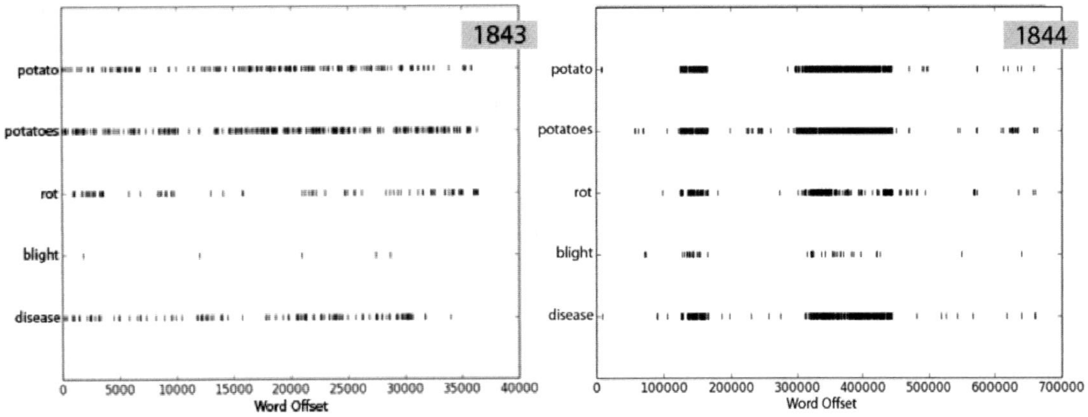

Figure 2: Lexical dispersion of key terms in the 1843 and 1844 patent reports.

particularly subject experts, can refine the results. A user can click on marked locations to view the linked passage. The user can also mark the passage to be accepted (if the information is valuable and the geographic location is of interest), moved (if the coordinates are incorrect), archived (if the information is valuable, but the geographic location is not relevant), or removed (if the information is of no value). Once the results are classified, they can then be used for spatial analysis (deliverable 5 in Table 1). This spatial analysis is outside the scope of this paper.

3. Results

3.1. Data summaries

NLTK lexical dispersion plots of 1843 and 1844 patent reports generated showing word frequency and location in the text indicate disease dynamics (Fig. 2). In 1843, it is evident that the terms potato and disease occur consistently throughout the text, with no obvious cluster of terms (Fig. 2, left). However, the dispersion plot from 1844 reveals a dense discussion of "potato" and "disease" about halfway through the report around the time when the disease report first appeared (Fig. 2, right). Also, the term rot begins to be used and is correlated with the mention of potato and disease. The strong likelihood the authors were talking about late blight is indicated since the disease is mentioned with the term rot. Since the pathogen had not been officially named *P. infestans* or the disease late blight at the time, those terms were not found.

Other terms that describe the symptoms of disease were used. We performed a keyword search on an expanded potato disease term list across nine years of patent reports. The results show a spike in 1844, there was a spike in the usage of the keywords "blight", "malady", "murrain", and "plague", with "blight" being the most frequent and plague used the least (Fig. 3).

We were able to learn more about how farmers were attempting to control the potato disease. In 1844, we found evidence that farmers were most commonly using lime as a means of control. One author noted: "To cleanse these cellars and whitewash them with lime are excellent means of destroying the germs" (Office 1841-1850). Other control methods included: pounded coal, salt, and sulphuric acid.

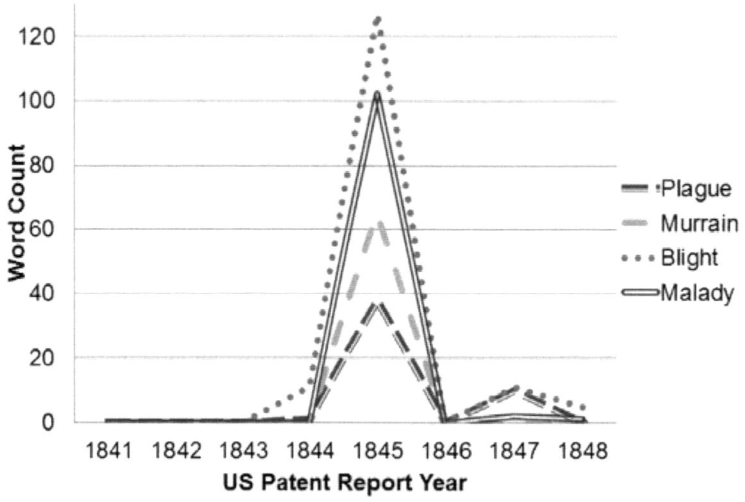

Figure 3: Keyword counts for late blight terms in reports from 1841-1848.

For our large corpus Google Books summary, we conducted an Google Ngram search on three 2-grams, late blight (red in Fig. 4), potato disease (green in Fig. 4), and potato rot (blue in Fig. 4). As suspected, the mention of potato disease spiked around 1845 and the mid-1870s. These time frames were known to be periods when serious *P. infestans* outbreaks occurred. The term "potato rot" began to be used shortly after early outbreaks in the 1840's. However, the term "late blight" began to be used around 1900, which is approximately 60 years after the Irish Potato Famine. By this time breeding work had led to the development of some cultivars that delayed onset of disease, thus the "late blight" as seen with the red line in Fig. 4.

Figure 4: Google Ngram Viewer for keywords: "late blight", "potato disease", "potato rot".

3.2. Mapped geoparsing results

Geoparsing was conducted on text extracted from the 1843 to 1848 annual patent reports with the KeyTermSearcher on the potato terms in the first row of Table 1. The distribution patterns of place names in these text passages are mapped in Fig. 5. In general, the locations were distributed mainly in northeastern US and western Europe.

Co-occurrence of potatoes and disease are mapped for the years 1844 and 1845 when the disease was rampant (Fig. 6). Locations associated with potato and disease terms (italicized in

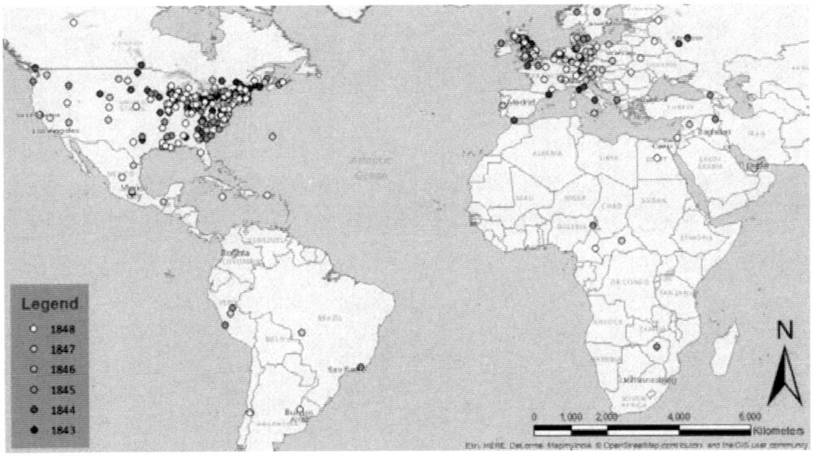

Figure 5: Place names mentioned near potato terms in the text from 1843-1848 patent reports.

Table 1, rows 1 and 2) are marked with circles and crosses, respectively. In both years we see an overlap of potato and disease discussions, clustering mainly in the northeast US and Europe. In 1845 we begin to see the mention of disease in South America, a region hypothesized to be a source of the disease in the US.

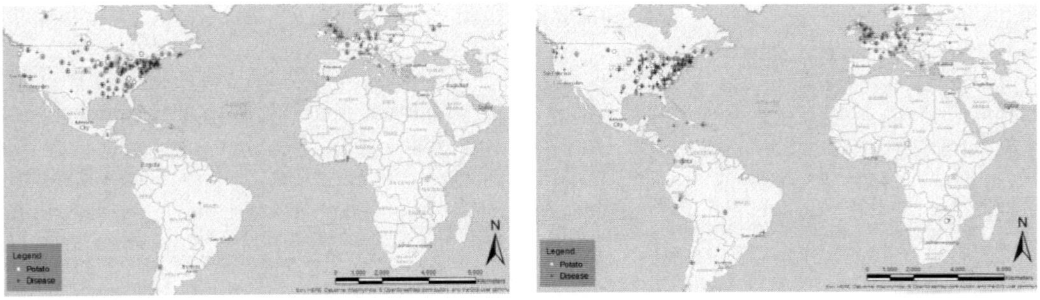

(a) (b)

Figure 6: Co-occurrence of potatoes and disease in 1844 (a) and 1845 (b) from maps created from US Patent Office Reports.

KeyTermSearch and geoparsing results were integrated into an interactive map. The results are mapped as selectable points that link to the text passages. Key terms in the text passages (e.g., potatoes) are linked to an HTML page which jumps the that text in the full text to provide further context for passages of interest. In the screenshot in Fig. 7, a user has selected a point in Kentucky which was mentioned as a place where potatoes crops were poorer in the previous year. The user can select to accept this data record (in the drop down menu), since it contains useful information. So this tool facilitates both data exploration and cleaning.

4. Discussion

NLTK summary techniques were useful for gaining preliminary familiarity with the patent report data. Concordances and co-locations helped us to refine our key search terms and list of terms to omit. The lexical dispersions and keyword frequency plots (e.g., Figs. 2-4) revealed change points in the data.

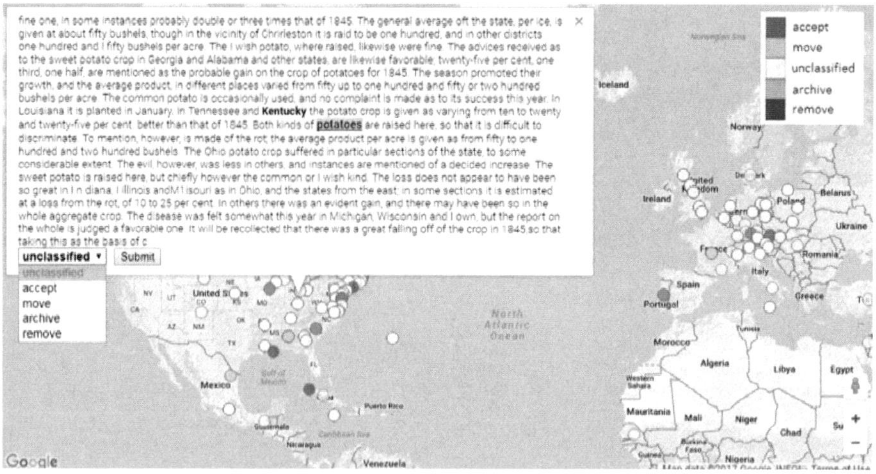

Figure 7: Interactive map for exploring text near potato terms.

Geoparsing and mapping text selections from time-series of historic passages combined with interactive map post-processing is a promising approach for tracking disease origin and spread patterns. Maps like Fig. 7 maps show regional clustering of potato and disease related terms. Modifications to the geoparsing tools to add spatially bounded searches could improve the results (e.g, no state west of the Mississippi allowed). Our key term search can be refined for the categories that are only of interest when located near a potato term.

The potato blight is one example of many in which context-based spatial searches may help experts to explore unstructured text and make discoveries. In this way, as we develop our methodologies, we hope to inform research in other domains. Scripts associated with this project are available in our github repository https://github.com/lgtateos/potato-text-mining.

5. Conclusions

We presented approaches for mining text documents and extracting geospatial information related to the historic spread of a crop disease that had severe consequences. These text analytic tools can provide useful information regarding the location and frequency of late blight in the 19th century. We were able to map the co-occurrence of potatoes and disease terms to identify locations of overlap. Next steps for extending this work will employ geospatial analytics to locate hotspot areas of late blight and the routes in which the pathogen traveled between locations. We plan to extend our text analysis to include contemporary news articles form the time. With newspapers, we can focus on known locations of interest, such as areas near the shipping ports in New York and Philadelphia, where the disease was initially introduced.

Our results show that text mining techniques, along with interactive mapping interfaces can augment expert analysis to support knowledge discovery in historic disease spread. This work can fill a critical gap in the history of late blight with historically accurate information that will help researchers worldwide understand the onset and spread of one of the most devastating plant pathogens. With the massive online news event data being collected in online databases, near real-time disease tracking and spread prevention may soon be possible with text parsing and mapping tools.

References

Birch, P. R. J., Cooke, D. E. L., 2013. The early days of late blight. eLife 2:e00954.

Bourke, P. M., 1964. Emergence of potato blight. Nature 203(4947), 184346,805–808.

DeBary, A., 1876. Researches into the nature of the potato-fungusphytophthora infestans. J Roy Agr Soc 12, 239268.

Gomez-Alpizar, L., Carbone, I., Ristaino, J. B., 2007. An andean origin of phytophthora infestans inferred from mitochondrial and nuclear gene genealogies. Proceedings of the National Academy of Sciences of the United States of America 104, 3306–3311.

Goss, E. M., Tabima, J., Cooke, D., Restrepo, S., Fry, W. E., Forbes, G. A., Fieland, V. J., Cardenas, M., Grunwaldg, N. J., 2014. The irish potato famine pathogen phytophthora infestans originated in central mexico rather than the andes. Proceedings of the National Academy of Sciences of the United States of America 111, 8791–8796.

Greenbacker, C., 2013. Open source software for geotagging unstructured big data - clavin. Berico Technologies. Presentation at All Things Open.

Jean-Baptiste, M., et. al., 2011. Quantitative analysis of culture using millions of digitized books. Science 331, 176–182.

Martin, M. D., Vieira, F. G., Ho, S. Y., Wales, N., Schubert, M., Seguin-Orlando, A., Ristaino, J. B., Gilbert, M. T., 2015. Genomic characterization of a south american phytophthora hybrid mandates reassessment of the geographic origins of phytophthora infestans. Mol. Biol. Evol. 33, 478–491.

Office, U. S. P., 1841-1850. Annual report of the commissioner of patents. Washington, DC.

Tateosian, L. G., Glatz, M., Shukunobe, M., Chopra, P., 2017. Gazegis: A gaze-based reading and dynamic geographic information system. In Eye Tracking and Visualization. Mathematics and Visualization, 129–147.

Yoshida, K., Schuenemann, V. J., Cano, L. M., Pais, M., Mishra, B. R. S., Lanz, R. C., Martin, F. N., Kamoun, S., Krause, J., Thines, M., Weigel, D., Burbano, H. A., 2013. The rise and fall of the phytophthora infestans lineage that triggered the irish potato famine. eLife 2:e00731, 1–25.

The Utility of Beautiful Visualizations

Laura Tateosian[a,*], Reza Amindarbari[a], Christopher Healey[b], Pavel Kozik[c], James Enns[c]

[a] *Center for Geospatial Analytics, NC State University, Raleigh, NC*

[b] *Department of Computer Science, NC State University, Raleigh, NC*

[c] *Psychology Department, University of British Columbia Vancouver, BC, Canada*

Abstract: Geovisualizations provide a means to inspect large complex multivariate datasets for information that would not otherwise be available with a tabular view or summary statistics alone. Aesthetically appealing visualizations can elicit prolonged exploration and encourage discovery. Creating data geovisualizations that are effective and beautiful is an important yet difficult challenge. Here we present a tool for rendering geovisualizations of continuous spatial data using impressionist painterly techniques. The techniques, which have been tested in controlled studies, vary the visual properties (e.g., hue, size, and tilt) of brush strokes to represent multiple data attributes simultaneously in each location. To demonstrate this technique, we render two examples: 1) weather data attributes (e.g., temperature, windspeed, atmospheric pressure) from the NOAA Global Forecast System and 2) fragile state indices as assessed by the Foreign Policy Magazine. These examples demonstrate how open source geospatial visualizations can harness aesthetics to enhance visual communication and viewer engagement.

*Corresponding author
Email address: lgtateos@ncsu.edu (Laura Tateosian)

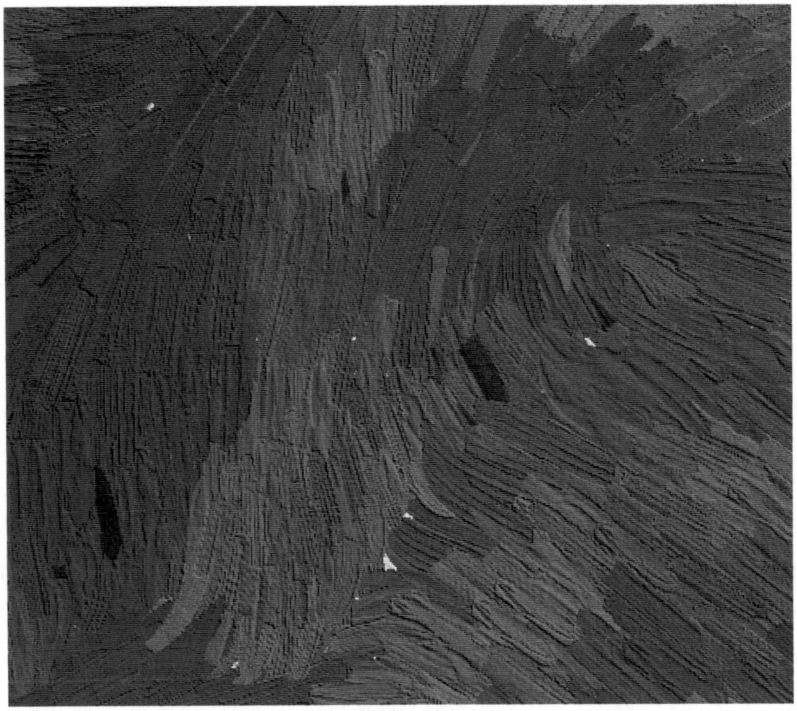

Figure 1: Painterly geovisualization of atmospheric data.

1. Introduction

As geospatial data collection and storage capabilities continue to grow, innovative visualization techniques are needed to support the extraction of information from datasets. Visualizations, graphical representations of data, can provide valuable insights into the datasets they represent. The term, visualization, is also used to refer to the process of converting raw data into images. Often, each of the data elements in a large data set has multiple attributes. For example, scientists studying the climatic El Niño data collect wind data, humidity, air temperature, sea surface temperature, latitude, and longitude, and other variables at buoys spanning the Pacific Ocean. Thus, in the resulting dataset an element represents a buoy and each element has more than six attributes. To visualize an m-dimensional data set of n elements, a set of visual features, V = $\{V_1, ..., V_m\}$ can be chosen to represent the data's set of attributes, A = $\{A_1, ..., A_m\}$. Examples of visual features include hue, texture, flicker, size, orientation, curvature, and regularity.

One approach to visualizing large multivariate datasets is glyph-based visualization Borgo et al. 2013. Glyphs are small visual elements, whose individual visual properties (color, shape, size, etc.) can be varied separately to represent the variations in the data attributes. Since glyphs can display several visual features at once, they can convey information about several attributes simultaneously. Studying human perception provides insights into how to use glyphs to best harness the human visual system. Human vision operates in two distinct modes Julesz 1984. First, preattentive vision is instantaneous and independent of the number of elements in the display and covers a large visual field. Preattentive vision includes the psychology concept called orienting, which refers to a reflexive direction of attention in response to a salient local stimulus, such as the sudden appearance of a new object Healey and Enns 2012. Second, attentive vision searches serially by focal attention and is limited to a small spatial aperture.

Attentive vision requires engagement, which means a viewer's attention is held for a longer period of time at each spatial locatio Healey and Enns 2012.

Geovisualizations that are designed for human perceptive strengths *and* aesthetic appeal may improve engagement and help to direct the viewer to key components of the representation, as the brush strokes in painterly visualization in Fig. 1 draw the viewer's eyes down and to the left conveying the dominant motion of the underlying flow data. Techniques from the realm of the visual arts can inform design decisions in geovisualization. Geovisualization can benefit from a long-standing interest in the computer graphics community in developing rendering algorithms for artistic techniques.

The computer graphics fields of non-photorealistic rendering (NPR) integrates artist techniques into graphics renderings with early examples simulating various artistic media, including pen-and-ink drawing, watercolor, charchoal, oil painting, and colored pencil. Strothotte and Schlechtweg 2002 provides a comprehensive review of this field. Ideas from NPR inspired non-photorealistic layered visualizations of brain matter Laidlaw et al. 1998 and airfoil wind flow patterns Kirby et al. 1999, as well as stippling Lu et al. 2002 and boundary enhancement Rheingans and Ebert 2001 for medical volume visualizations. Others began to bring NPR into geovisualization with natural patterns on agricultural maps Interrante 2000 and bus-maps with Mondrian painting designs Skog et al. 2003. Recent work developed a direct-manipulation user interface to enable artists to design new data geovisualizations Schroeder and Keefe 2016. The resulting geovisualizations exhibit a masterful use of color. Also, web designers are developing appealing geovisualizations that consume global datasets[1] and near real-time weather data Beccario 2014.

In this paper, we look at an example of aesthetic geovisualization that is inspired by Impressionist painter, Vincent Van Gogh. Visual features of brush-stroke glyphs portray multiple data attributes at each position. Tateosian et al. 2007 describes the design and algorithms of these techniques. The painting algorithm randomly selects unpainted data positions for brush strokes until the desired coverage is reached and assigns visual features, stroke color, size, and tilt, to represent the underlying data at the center of its assigned position.

Tateosian et al. 2007 developed three distinct visualization styles based on properties of aesthetics. The examples in this paper feature *visual complexity*, a style resembling the later works of Vincent van Gogh. In work by Kozik et al. accepted the three painterly styles, along with a standard visualization employing rectangular glyphs on a regular grid were shown to participants in a set of experiments. A first experiment found that visual complexity was comparable to standard glyphs and superior to the other painterly styles for memorability. A second experiment tested participants' ability to report the prevalence of the color blue (representing a single weather condition) within each visualization. For this task, visual complexity was superior to glyphs and the other impressionist styles.

The implementation of the visualization techniques demonstrated here are available in a C++/OpenGL painterly visualization code base[2]. The visual complexity style uses a fast paint texture NPR technique Hertzmann 2002 to simulate a painted surface by assigning height values to texture maps on the strokes and calculating the lighting based on those values. The user can assign six data attributes to six visual features. The first four features are the hue, luminance, size, and tilt of brush strokes. The last two are termed *contrast* and *proportion*. Contrast changes the tilt and luminance of a subset of strokes in a region. In Fig. 2a a few dark green strokes at a 45°tilt contrast the field of light green tilted at 135°. The tilt of the contrast

[1] https://www.jasondavies.com/

[2] NPV code available at https://github.com/lgtateos/pittore-rep

(a) Contrast strokes (b) Proportion (long+narrow) (c) Proportion (short+wide)

Figure 2: Visual features, *contrast* and *proportion*, extend the standard visual lexicon.

strokes is rotated by 90 degrees and the luminance is shifted down to create noticeably darker strokes perpendicular to the dominant stroke direction in that area. The proportion attribute uses three bins, high, very high, and the rest. The standard proportion for strokes is 1:4 (one wide to four long). When the value of the attribute assigned to proportion exceeds a very high threshold, the stroke length is exaggerated and the width is narrowed, changing the proportion to 1:25. In Fig. 2b, light pink long narrow strokes are centered on values which fall above the threshold. High, but not very high, are represented with wide short strokes (7:5 ratio). Peach-colored strokes in the upper right corner and orange strokes in the upper left corner of Fig. 2c are centered on high values. The direction of the stroke is still apparent because of the simulated bristle texture. Visual complexity style also implements b-splines for stroke curvature that can be used for flow visualization (Fig. 1).

Computational geometry techniques are used to segment the data into regions based on the set of underlying data values at each point. Each region is painted by first randomly selecting an unpainted pixel, sizing and tilting the stroke based on underlying data, and then checking the segment boundaries and the overlap with existing strokes. If the latter are within acceptable thresholds, the stroke is placed there. This continues until the desired segment coverage is met. Tateosian et al. 2007 describes the segmenting and painting algorithms in detail. This stand-alone code uses a Mercator projection for stroke placement as it has not yet been integrated with existing geographic libraries. Examples of practical applications for painterly and ideas for integrating aesthetics into geovisualization tools to increase expressiveness are presented in the following section.

2. Painterly geovisualizations

Here we test the visual complexity style geovisualizations on large multivariate datasets within two domain applications. The first application, atmospheric data, is a spatially continuous dataset, sampled at regular intervals. The second domain, fragile states indices, portrays polygonal data, with continuous attributes assigned to each country, instead of regular grid cells.

2.1. Weather data time series

The University Corporation for Atmospheric Research (UCAR) hosts massive volumes of atmospheric data. These datasets enable atmospheric scientists to study processes that can lead to hurricanes or cyclones, such as El Niño and African Easterly Waves, problems that entail inspecting combinations of attributes such as wind direction and water surface temperature. We extracted NOAA Global Forecast System data collected at 0.25° resolution and six-hour intervals[3]. Six of the available attributes were selected to make up these datasets, temperature,

[3] See http://thredds.ucar.edu/ for Global Forecast System data

geopotential height, and pressure at maximum wind level, precipitable water, and finally, u and v components of wind at maximum wind level, which were combined to calculate wind direction.

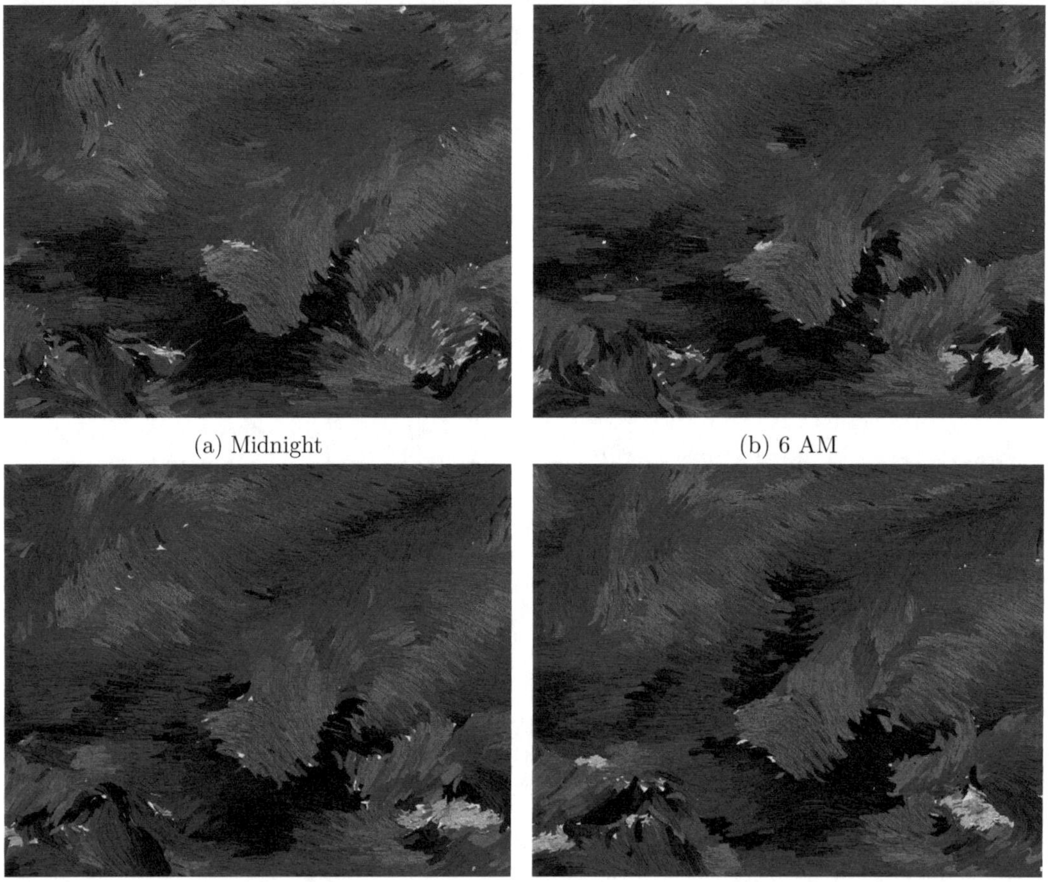

(a) Midnight (b) 6 AM

(c) Noon (d) 6 PM

Figure 3: Painterly geovisualization of atmospheric data mapping: temperature → stroke hue, geopotential height → stroke size, wind direction → stroke tilt, pressure → contrast strokes, cloud water → stroke proportion.

Figs. 3a-3d show four time-steps from the same day (May 25, 2017), six hours apart starting at midnight. Temperature, geopotential height, and wind direction are represented with stroke hue (blue-green-brown-orange-pink), size (small to large), and tilt. Pressure and cloud water are represented with contrast strokes and stroke proportion. Studying these images, we see several distinctive features: shifting turbulence near the lower right corner, warmer temperatures with a high pressure system appearing in the upper left at 6pm, and lower geopotential height in the north than the south all day.

2.2. Fragile state index

The Foreign Policy Magazine releases an annual Fragile States Index, rating each country on a number of measures to model the country's fragility or likelihood to become a failed state. A committee of economists assigns values for indices on a scale from 1 for good to 10 for poor. Twelve factors are evaluated and summed to compute a total which is used to rank order 178 countries in terms of fragility. Fig. 4 visualizes the overall state fragility ranking and four of the ranking factors, including poverty and economic decline, protection of human rights, demographic pressures such as food scarcity and population growth, and state legitimacy as a measure of corruption, government performance, and the electoral process. We retrieved the data from the Foreign Policy Magazine web site[4].

The rank ordering in Fig. 4 is mapped to the majority color of the brush strokes, which varies from low to high on the perceptually balanced continuous blue-green-brown-orange-pink color ramp, e.g., pink central African countries are highly fragile, while blue Canada has low fragility. poverty and economic decline is mapped to stroke size, e.g., Guinea in northwest Africa has large strokes due to the high poverty rates there. In contrast, states like Norway have very small strokes to represent their very low poverty rates. Protection of human rights is mapped to stroke tilt, with the vertical stroke in central Africa reflecting human rights concerns and the horizontal strokes in New Zealand on the other end of the spectrum. Demographic pressure is mapped to contrast strokes. Strokes that contrast in luminance and tilt to the color and tilt assigned for that state represent those above a threshold. For example, China and Brazil have high population growth pressures while Australia doesn't, since the strokes are monotone and all tilt in the same direction. State legitimacy is mapped to stroke proportion, which overrides the standard proportionality (1:4) of the strokes. When corruption values are very high the strokes are long and narrow relative to the standard proportion. For example, North Korea and the Congo are painted with a number of long narrow strokes, since they have poor state legitimacy ratings. Corruption in Egypt falls within the high (but not very high) range with short wide strokes visible there. The tool provides interactive zoom and pan with the mouse for users to inspect the visualization more closely.

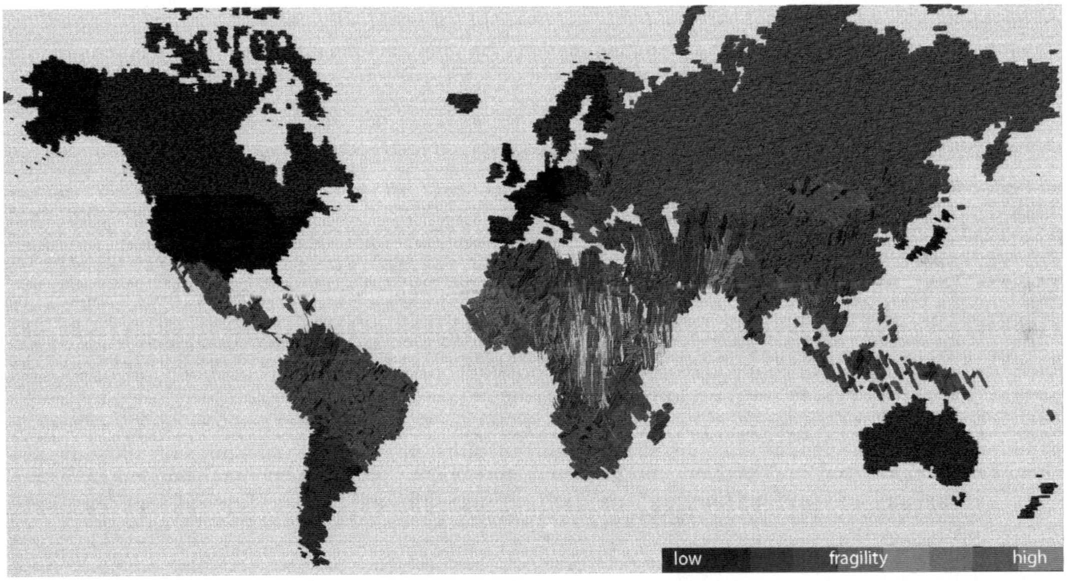

Figure 4: Painterly geovisualization of the 2016 state fragility index by the Foreign Policy magazine (FSI Data, 2017). Mapping fragility → stroke hue, poverty and economic performance → stroke size, human rights to stroke tilt, contrast strokes → demographic pressure, stroke proportion → state legitimacy (corruption).

3. Conclusions

The glyph-based approach enables multiple attributes to be visualized in a single location, so that interactions may be determined and appealing geovisualizations such as the visual complexity examples shown here may engage viewers and result in new discoveries. GIS platforms

[4]http://foreignpolicy.com/fragile-states-index-2016-brexit-syria-refugee-europe-anti-migrant-boko-haram/

(both open source and proprietary) that can support artistic geovisualizations are almost non-existent. Integrating stand-alone tools such as this into existing open source GIS systems will further enhance the visualizations. Established geospatial frameworks can provide the capability to apply different projections, to overlay the images with other spatial layers for geospatial context, generate legends, provide graphical user interfaces to guide users and add picking for inspecting individual elements. At the same time, geovisualization developers may be able to develop tools that make it easier to develop more expressive and engaging visualizations by leveraging the creativity and visual salience of artistic techniques.

References

Beccario, C., 2014. Building earth: an animated map of global wind and weather. Presentation at The Graphical Web: Visual Storytelling, Winchester, Hampshire, UK. August 26-30.

Borgo, R., Kehrer, J., Chung, D. H., Maguire, E., Laramee, R. S., Hauser, H., Ward, M., Chen, M., 2013. Glyph-based visualization: Foundations, design guidelines, techniques and applications. In: Eurographics (STARs). pp. 39–63.

Healey, C., Enns, J., 2012. Attention and visual memory in visualization and computer graphics. IEEE transactions on visualization and computer graphics 18 (7), 1170–1188.

Hertzmann, A., 2002. Fast paint texture. In: Proceedings of the 2nd international symposium on Non-photorealistic animation and rendering. ACM, pp. 91–ff.

Interrante, V., 2000. Harnessing natural textures for multivariate visualization. IEEE Computer Graphics and Applications 20 (6), 6–11.

Julesz, B., 1984. A brief outline of the texton theory of human vision. Trends in Neurosciences 7 (2), 41–45.

Kirby, R. M., Marmanis, H., Laidlaw, D. H., 1999. Visualizing multivalued data from 2d incompressible flows using concepts from painting. In: Proceedings of the conference on Visualization'99: celebrating ten years. IEEE Computer Society Press, pp. 333–340.

Kozik, P., Tateosian, L., Healey, C. G., Enns, J. T., accepted. Impressionism-inspired data visualizations are both functional and beautiful. In: Psychology of Aesthetics, Creativity, and the Arts.

Laidlaw, D. H., Ahrens, E. T., Kremers, D., Avalos, M. J., Jacobs, R. E., Readhead, C., 1998. Visualizing diffusion tensor images of the mouse spinal cord. In: Visualization'98. Proceedings. IEEE, pp. 127–134.

Lu, A., Morris, C. J., Ebert, D. S., Rheingans, P., Hansen, C., 2002. Non-photorealistic volume rendering using stippling techniques. In: Visualization, 2002. VIS 2002. IEEE. IEEE, pp. 211–218.

Rheingans, P., Ebert, D., 2001. Volume illustration: Nonphotorealistic rendering of volume models. IEEE Transactions on Visualization and Computer Graphics 7 (3), 253–264.

Schroeder, D., Keefe, D. F., 2016. Visualization-by-sketching: An artist's interface for creating multivariate time-varying data visualizations. IEEE transactions on visualization and computer graphics 22 (1), 877–885.

Skog, T., Ljungblad, S., Holmquist, L. E., 2003. Between aesthetics and utility: designing ambient information visualizations. In: Information Visualization, 2003. INFOVIS 2003. IEEE Symposium on. IEEE, pp. 233–240.

Strothotte, T., Schlechtweg, S., 2002. Non-photorealistic computer graphics: modeling, rendering, and animation. Morgan Kaufmann.

Tateosian, L. G., Healey, C. G., Enns, J. T., 2007. Engaging viewers through nonphotorealistic visualizations. In: Proceedings of the 5th international symposium on Non-photorealistic animation and rendering. ACM, pp. 93–102.

Urban Multi-scale Environmental Predictor - an extensive tool for climate services in urban areas

Fredrik Lindberg[a,*], Sue Grimmond[b], Andrew Gabey[b], Bei Huang[f,b], Christoph W Kent[b], Ting Sun[b], Natalie E Theeuwes[b], Leena Jarvi[c], Helen Ward[b], Izzy Capel-Timms[b], YY Chang[e], Per Jonsson[d], Niklas Krave[a], Dongwei Liu[e], D Meyer[b], K Frans G Olofson[a], Jian-Guo Tan[e], Dag Wastberg[d], Lingbo Xue[b], Zhe Zhang[b]

[a]*Department of Earth Sciences, University of Gothenburg, Gothenburg, Sweden*

[b]*Department of Meteorology, University of Reading, Reading, United Kingdom*

[c]*Department of Physics, University of Helsinki, Finland*

[d]*Tyrns AB, Gothenburg, Sweden*

[e]*Shanghai Institute of Meteorological Science, Shanghai Meteorological Service, China*

[f]*Tsinghua University, Beijing, China*

Abstract: The city based climate service tool UMEP (Urban Multi-scale Environmental Predictor) is a coupled modelling system that combines models essential for urban climate processes and is developed as an extensive QGIS plugin. An application is presented to illustrate its potential, specifically of the identification of heat waves and cold waves in cities. The tool has broad utility for applications related to outdoor thermal comfort, urban energy consumption, climate change mitigation etc. It includes tools to: enable users to input atmospheric and surface data from multiple sources, prepare meteorological data for use in urban areas, undertake simulations and consider scenarios, and compare and visualize different combinations of climate indicators.

*Corresponding author
Email address: `fredrikl@gvc.gu.se` (Fredrik Lindberg)

1. Introduction

Scientists and practitioners from disciplines, such as architecture (e.g. Ren et al. 2011), climatology (e.g. Eliasson 2000), planning (e.g. Alcoforado et al. 2009), engineering and geography, have long been interested in how weather and climate affects cities and city dwellers. However, city based climate services, which require climate knowledge and information, for urban planning, building design and the operation of cities is not straight-forward (Chrysoulakis et al. 2013; Baklanov et al. 2017; Grimmond et al. 2014). Models and tools at the appropriate scales are only now being developed. Input data (e.g. surface and atmospheric) can be challenging to access and/or in formats that are inaccessible to many end-users. Communication between producers and users of climate services has been poor, with outputs that can be difficult to interpret by practitioners and non-specialists. These challenges indicate a need for tools that are more user-friendly, and technically and economically accessible to improve communication across disciplines, researchers and practitioners.

To address this need, we introduce UMEP (Urban Multi-scale Environmental Predictor) an integrated tool for urban climatology and climate-sensitive planning applications. This tool can be used for applications related to outdoor thermal comfort, solar energy production, climate change mitigation etc. UMEP consists of a coupled modelling system which combines different kinds of models with systems to input data from multiple sources, formats and at different temporal and spatial scales, and to generate output as data, graphs and/or maps. An important feature of UMEP is its ability to address issues and relevant processes at the different temporal and spatial scales investigated. Here the structure of UMEP and a practical application to illustrate its potential are presented.

UMEP is being developed as a community tool using only open source software and libraries to enable its use without restriction to users and researchers. Users may contribute to the tool to enhance and extend its capabilities. A major feature is the coupling of climate data with spatial information to determine model parameters, and to edit, map and visualize inputs and results. Therefore, the software is written as a plug-in to QGIS, a cross-platform, free and open source desktop geographic information system (GIS) application (Team 2016).

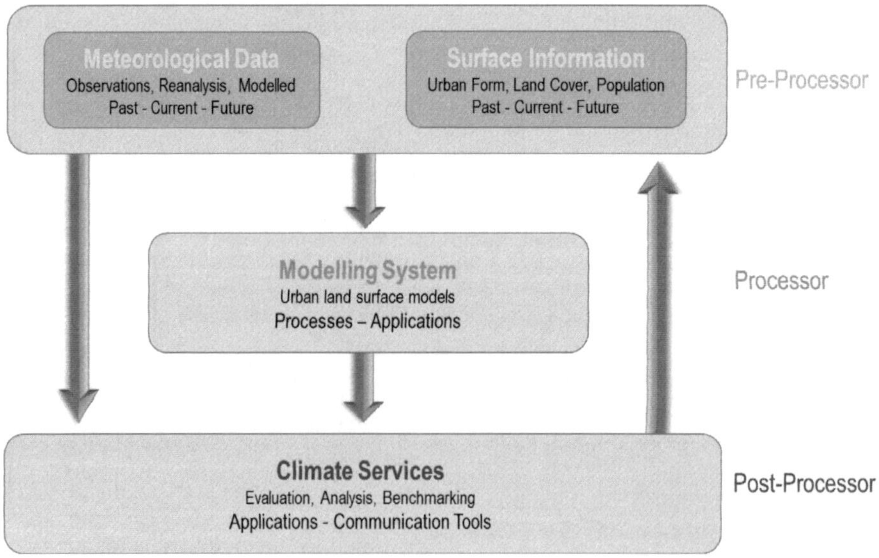

Figure 1: Structure of UMEP (Urban Multi-scale Environmental Predictor).

UMEP has three main elements (Figure 1): pre-processor (for inputs of meteorological and surface information); processor (modelling system e.g. Urban Land Surface Models, ULSM); and post-processor (tools to analyze the outputs (individual and ensemble, indicators of uncertainty, user applications etc.). Each separate tool (plug-in) is described briefly in Table 1, with full details presented in the online manual (www.urban-climate.net/ umep/UMEP). UMEP allows users to: integrate atmospheric and surface data from multiple sources; take data measured at 'standard' sites and adapt it to be representative location within the urban environment; and compare and visualize results or scenarios for different climate indicators of concern or interest (heat indices, intense precipitation, water/energy demand) at a range of spatial scales consistent with end-users' needs and interests. For ease of use, tutorials have been developed (http://www.urban-climate.net/umep/UMEP_Manual Tutorials) for users to try out the different parts of the tool. A key contribution of UMEP is to facilitate the preparation of input data needed for City-Based Climate Services (CBCS). UMEP provides both guidance and tools that enable the preparation and manipulation of data. These tools can be found in the pre-processor section (Table 1). This is particularly important as most end-users are familiar with some, but not the full spectrum of data needed for applications. For example, planners are knowledgeable about buildings heights, materials and their spatial arrangement (i.e. urban surface data) and often have geographic information system (GIS) skills, but they do not necessarily have detailed knowledge of meteorological data.

2. Structure and Capabilities of UMEP

UMEP broad range of capabilities includes tools that may be used independently and/or in combinations depending on the application. The output from pre-processor tools may be used for other external modelling applications, or within a chain of tools to provide climate indicators for decision making. Many of the individual tools, and their evaluation, have been described in numerous previous papers; see Lindberg et al. 2017 for details.

3. Application Example - Energy and Water Balance Fluxes

Energy and water balance fluxes are critical to surface-atmosphere interactions in an urban area. The Surface Urban Energy and Water balance Scheme (SUEWS) is a core urban land-surface model included in the processing part of UMEP (Table 1). The model is able to simulate the urban radiation, energy and water balances using commonly measured meteorological variables and information about the surface cover. SUEWS is applicable at the local (neighborhood) to city scale. UMEP has the latest version of SUEWS (Ward and Grimmond 2017, Ward et al. 2016) accessible through two links:

(a) SUEWS Simple: is intended to provide a useful starting place to introduce UMEP and SUEWS. Example data are provided so that the user can explore the impact of modifying urban surface characteristics. With SUEWS Simple, the ULSM can be executed for a single location (area).

(b) SUEWS (Advanced) provides a full version of the model appropriate for both spatial and temporal investigation of the urban energy balance.

The SUEWS model has been extensively evaluated at various locations and situations worldwide (see Lindberg et al. 2017).

Component	Scales	Description
Pre-Processor		
Prepare existing data	C/L/M	Formats meteorological data for input to the models, dealing with missing data
Download data (WATCH)	C/L	Re-analysis data modified for use in an urban context, representative of the local scale
Spatial Data Downloader	C/L/M	Downloads spatial data from public servers
Tree Generator	C/L/M	Creation/manipulation of 3D vegetation data
LCZ Converter	C/L	Allows morphometric parameters and land cover fractions to be calculated from Local Climate Zone (LCZ) maps.
Sky View Factor	L/M	Amount of the hemisphere with restricted view of the sky
Wall Height Aspect	L/M	Height and orientation of buildings and walls
Land cover reclassifier	C/L/M	Geodata can be translated into the land cover classes used by all the models
Land cover fraction (point)	L/M	Surface cover fractions are determined for an area or specific directions
Land cover fraction (grid)	C/L	As above, but a grid is used to determine fractions for multiple areas.
Morphometric Calculator (point)	L/M	Morphometric parameters are determined for an area or specific directions can be used.
Morphometric Calculator (grid)	C/L	As above, but a grid is used to determine parameters for multiple areas.
Source area (point)	L/M	As above, but determined for an area derived from source area models.
SUEWS Prepare	C/L	Prepares input data for the SUEWS model (processor).
Processor		
ExtremeFinder	C	Heat and cold wave conditions.
Mean Radiant Temperature (SOLWEIG)	L/M	SOLWEIG estimates spatial (2-D) variations of 3-D radiation fluxes and the mean radiant temperature (T_{mrt}) in complex urban settings.
Anthropogenic Heat (LQF)	C/L	Globally applicable method (low spatial resolution) to calculate Q_F
Anthropogenic Heat (GQF)	C/L	Locally applicable method (high spatial resolution) to calculate Q_F
Urban Energy and Water Balance (*SUEWS; Simple*)	C/L	Urban land surface model of radiation, energy and water fluxes for a single point or area.
Urban Energy and Water Balance (*SUEWS; Advanced*)	C/L	As above, but for multiple areas.
Solar Energy on Building Envelopes	L/M	Shortwave irradiance on ground, roofs and building walls.
Daily Shadow Patterns	L/M	Shadow maps are derived from buildings and 3D vegetation.
Post-Processor		
SEBE (Visualization)	L/M	Visualization of solar irradiation on building envelopes i.e., both roofs and walls.
SOLWEIG Analyzer	L/M	Analyzing output from the SOLWEIG model.
SUEWS Analyzer	C/L	Analyzing output from the SUEWS model.
Benchmarking	C/L/M	Statistical tool to compare different datasets

Table 1: Description of UMEP components and scales applicable to: C: city; L: local (neighborhood); M: micro (e.g. street canyon, park). Micro scale applications can be used across a whole city, but will likely be computer intensive.

3.1. Energy and water balance fluxes for Beijing, China

In this example the implication of a heat wave event compared to a 'normal' summer month is compared with regards to sensible heat flux over Beijing, China. The workflow for this application is outlined in Figure 2. Geo-datasets that contain information about the urban environment are used with the pre-processor tools to provide the required surface parameters. It is strongly recommended that all geodata used are transformed into the same projected coordinate system. The model can be applied to areas of any shape, here referred to as 'grids'. These areas need to be defined in a vector polygon layer. In most cities, there will be planning units with known boundaries already available in this format (e.g. boroughs, wards). In this example a square grid is created in QGIS, e.g. by using Vector >Research Tools >Vector grid.

As land cover information is a key characteristic for many calculations, a method to reclassify data is provided (Land Cover Reclassifier). In this application example, no specific spatial input data exists and thus other sources of information is required. Accessing reliable sources of land cover information to derive these parameters at the scale of interest remains a challenge. Crowd-sourced data sets such as OpenStreetMap (http://www.openstreetmap.org) and WUDAPT (http://www.wudapt.org/) offer potential but may be incomplete or inconsistent. In this example, Local Climate Zones (Stewart and Oke 2012) is derived using the LCZ converter available from the UMEP pre-processor. This tool converts WUDAPT data into input information for SUEWS (Figure 3).

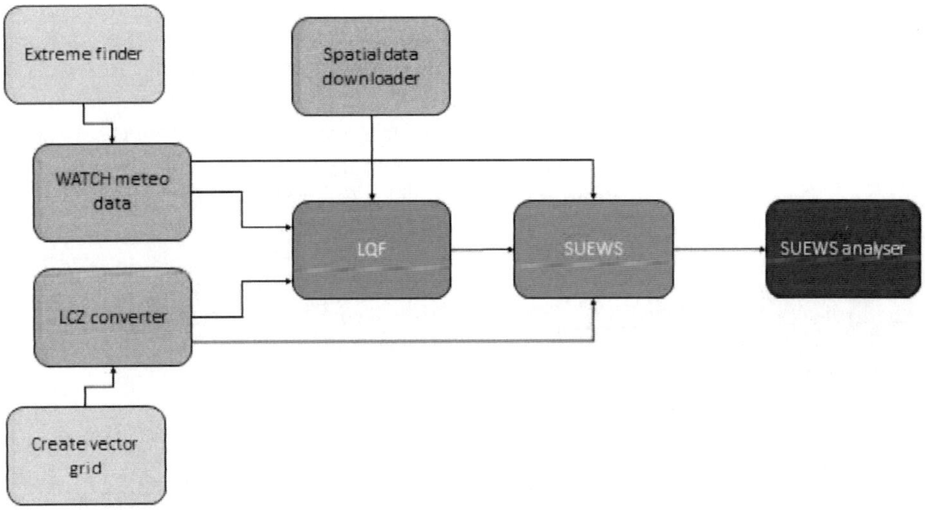

Figure 2: Workflow of an urban energy and water balance application.

Morphometric parameters required as model input can be calculated using the Morphometric Calculator (Grid) using digital surface models (DSM). From these data, the zero-plane displacement height (zd), aerodynamic roughness length (z0) and other geometric parameters such as mean roughness-element height and frontal area index can be calculated. In this application example this information is parameterized from the WUDAPT LCZs. However, if more detailed information is available, this can be utilized.

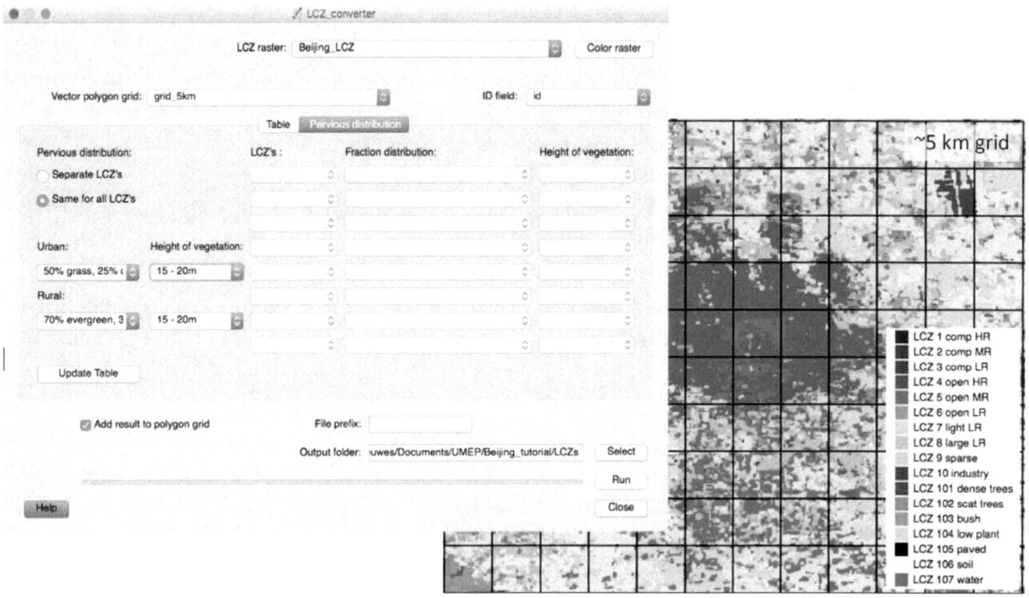

Figure 3: (left) LCZ Converter interface in UMEP and (right) Beijing LCZs derived from the WUDAPT database (www.wudapt.org).

Population density (people per hectare) is used in the estimation of anthropogenic heat flux in SUEWS. In this example population density datasets are unavailable (e.g. as would be obtained from local census data), so the Spatial Data Downloader can be used. This plug-in is directly connected to various Web Coverage Services (WCS) including global datasets on population density. Population information has to be (dis-)aggregated based on the polygon grid. This is preferably done using the built-in tool Zonal statistics in QGIS.

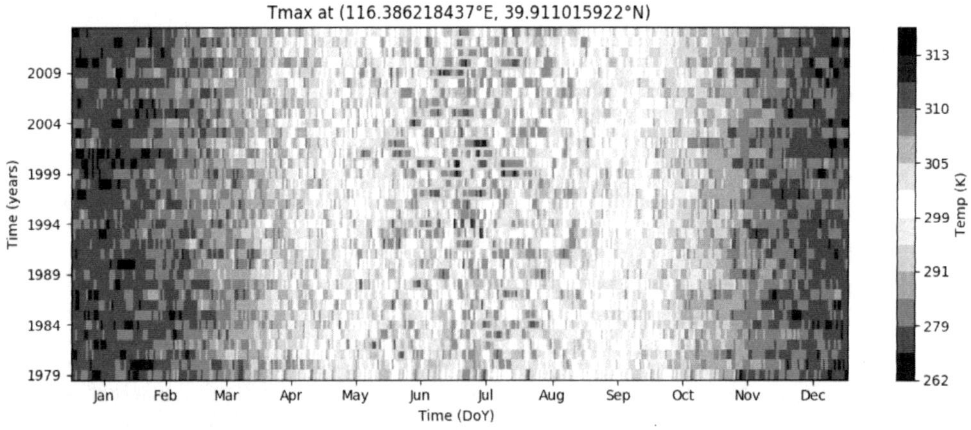

Figure 4: Output from the ExtremeFinder where yellow boxes are identified as heat wave periods.

The other major input is the forcing meteorological data. This needs to be for above the height of the roughness elements (trees, buildings). A common format is used in all UMEP models but unnecessary data for a specific application does not need to be supplied. Most applications require a continuous gap-filled data set. For many urban applications, the impact of daylight savings starting and finishing is important. In this example, the ExtremeFinder is used to identify years with 'normal' and 'heat wave' summer situations for Beijing (Figure 4). In this example 2006 was identified as a 'normal' and 2009 as a 'heat wave' summer. When the years of interest have been identified, the WATCH data plugin in UMEP is used to retrieve data from the years of interest. UMEP allows the user to draw on the reanalysis dataset, WATCH Forcing Data ERA-Interim (WFDEI) (Weedon et al. 2011). This product was selected as it was designed to be used for hydrological and land-surface modelling for climate purposes. It is derived from the ERA-Interim reanalysis product (Dee et al. 2011) via sequential interpolation to half-degree spatial resolution with 3 h temporal resolution (Weedon et al. 2011, Weedon et al. 2014).

Once all the required information has been pre-processed, the SUEWS model can be executed through the UMEP processor. Results from two example model runs are shown in Figure 5 where the increase in sensible heat (QH) is evident. The two maps are generated using the SUEWS Analyzer in the post-processor section in UMEP. This enables simple exploration of spatial and temporal model results.

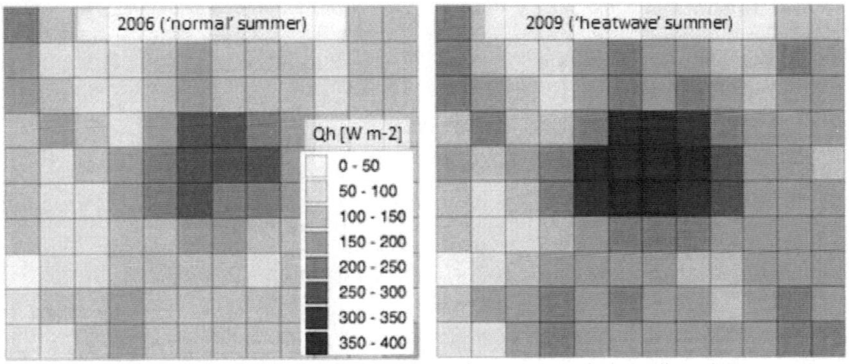

Figure 5: Mean sensible heat flux at 13:00 local time during July for Beijing, China.

4. Summary and Future Prospects

This paper introduces the Urban Multi-scale Environmental Predictor, UMEP and gives an example its potential through an application. Designed as a plug-in for QGIS, UMEP consists of a coupled modelling system which combines state of the art 1-D and 2-D models related to processes essential for urban climate estimation. The tool has broad utility for applications related to outdoor thermal comfort, urban energy consumption, climate change mitigation etc. The grid-based source area estimations of morphometric parameters such as z0 and zd (Kent et al. 2017) allows interpretation of observations and preparation of data for climate modelling. Planned developments include tools for pedestrian wind and thermal comfort indices; generation of site-specific forcing data using a coupled convective boundary layer (CBL) and land surface model (Onomura et al. 2015).

References

Alcoforado, M. J., Andrade, H., Lopes, A., Vasconcelos, J., 2009. Application of climatic guidelines to urban planning: The example of lisbon (portugal). Landscape and Urban Planning 90(12), 56–65.

Baklanov, A., Grimmond, C., Carlson, D., Terblanche, D., Tang, X., Bouchet, V., Lee, B., Langendijk, G., Kolli, R. K., 2017. Hovsepyan from urban meteorology, climate and environment research to integrated city services. Urban Climate.

Chrysoulakis, N., Lopes, M., Jose, R. S., Grimmond, C. S. B., Jones, M. B., Magliulo, V., Klostermann, J. E. M., Synnefa, A., Mitraka, Z., Castro, E. A., Gonzalez, A., Vogt, R., Vesala, T., Spano, D., Pigeon, G., Freer-Smith, P., Staszewski, T., Hodges, N., Mills, G., Cartalis, C., 2013. Sustainable urban metabolism as a link between bio-physical sciences and urban planning: The bridge project. Landscape and Urban Planning 112, 100 117.

Dee, D. P., Uppala, S. M., Simmons, A. J., Berrisford, P., Poli, P., Kobayashi, S., Andrae, U., Balmaseda, M. A., Balsamo, G., Bauer, P., Bechtold, P., Beljarrs, A. C. M., van de Berg, L., Bidlot, J., Bormann, N., Delsol, C., Dragani, R., Fuentes, M., Geer, A. J., Haimberger, L., Healy, S. B., Hersbach, H., Holm, E. V., Isaksen, L., Kallberg, P., Kohler, M., Matricardi, M., McNally, A. P., Monge-Sanz, B. M., Morcrette, J. J., Park, B. K., Peubey, C., de Rosnay, P., Tavolato, C., Thepaut, J. N., Vitart, F., 2011. The era-interim reanalysis: Configuration and performance of the data assimilation system. Quarterly Journal of the Royal Meteorological Society 137, 553597.

Eliasson, I., 2000. The use of climate knowledge in urban planning. Urban Planning 48, 3144.

Grimmond, S., Tang, X. U., Baklanov, A., 2014. Towards integrated urban weather, environment and climate services. WMO Bulletin 63(1), 10–14.

Kent, C. W., Grimmond, C. S. B., Barlow, J., Gatey, D., Kotthaus, S., Lindberg, F., Halios, C. H., 2017. Evaluation of urban local-scale aerodynamic parameters: implications for the vertical profile of wind and source areas. Boundary Layer Meteorology.

Lindberg, F., Grimmond, C. S. B., Gabey, A., Huang, B., Kent, C. W., Sun, T., Theeuwes, N., Jarvi, L., Ward, H., Capel-Timms, I., Chang, Y. Y., Jonsson, P., Krave, N., Liu, D., Meyer, D., Olofson, F., Tan, J. G., Wastberg, D., Xue, L., Zhang, Z., 2017. Umep - an integrated tool for city-based climate services. Environmental Modelling and Software.

Onomura, S., Grimmond, C. S. B., Lindberg, F., Holmer, B., Thorsson, S., 2015. Meteorological forcing data for urban outdoor thermal comfort models from a coupled convective boundary layer and surface energy balance scheme. Urban Climate 11, 1–23.

Ren, C., y. Ng, E. Y., Katzschner, L., 2011. Urban climatic map studies: a review. International Journal Of Climatology 31(15), 2213–2233.

Stewart, I. D., Oke, T. R., 2012. Local climate zones for urban temperature studies. Bulletin of the American Meteorological Society 93(12), 1879–1900.

Team, Q. D., 2016. Qgis geographic information system. open source geospatial foundation project. Online, online; accessed May 31, 2017.
URL `http://www.qgis.org/`

Ward, H. C., Grimmond, C. S. B., 2017. Using biophysical modelling to assess the impact of various scenarios on summertime urban climate across greater london. Landscape and Urban Planning.

Ward, H. C., Jarvi, L., Onomura, S., Lindberg, F., Grimmond, C. S. B., 2016. Suews manual: Version 2016a . Online, online; accessed May 31, 2017.
URL `http://www.urban-climate.net/umep/SUEWS`

Weedon, G. P., Balsamo, G., Bellouin, N., Gomes, S., Best, M., Viterbo, P., 2014. The wfdei meteorological forcing data set: Watch forcing data methodology applied to era-interim reanalysis data. Water Resources Research 50, 7505–7514.

Weedon, G. P., Gomes, S., Viterbo, P., Shuttleworth, W. J., Blyth, E., Osterle, H., Adam, J. C., Bellouin, N., Boucher, O., Best, M., 2011. Creation of the watch forcing data and its use to assess global and regional reference crop evaporation over land during the twentieth century. J. Hydrometerol 12, 823–848.

Posters

Download full posters from:
http://scholarworks.umass.edu/foss4g/vol17/iss1

Evaluation of the Micro-Tasking Method for OpenStreetMap Imports

Atle Frenvik Sveen[a,*], Anne Sofie Strand Erichsen[a]

[a]*NTNU/Norkart*

Abstract: Open Geospatial Data, capable of enriching OpenStreetMap, is being released by governments around the world at an increasing rate. The OSM import methods have been refined since the massive TIGER-import, moving towards assisted methods such as the with micro-tasking method used by the LA and NY buildings imports. While these imports serve as great case studies of imports, they do not deal with complex datasets, or updates to the data, neither do they deal with partitioning of tasks. We examine how the Norwegian FKB-dataset can be imported to OSM using micro-tasking, and perform a user-test to determine the best partition of these micro-tasking tasks. In this we hope to improve the micro-tasking method to enable efficient and correct imports and updates of Open Governmental Geospatial Data to OSM.

Poster Download: http://scholarworks.umass.edu/foss4g/vol17/iss1/20

*Corresponding author
Email address: `atlsve@norkart.no` (Atle Frenvik Sveen)

Evaluation of the Micro-Tasking Method for OpenStreetMap Imports

The problem

How to import detailed (building) data to OSM?

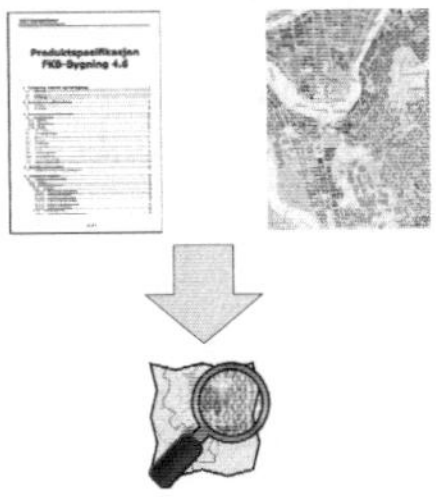

Import Strategies

Level of human involvement

- Fully automated
- Guided automated (Kartverket)
- Micro-Tasking (NYC)

Micro-Tasking

"Computers are incredibly fast, accurate, but stupid. Humans are incredibly slow, inaccurate, but brilliant. Together they may be powerful beyond imagination" – A. Einstein

- Split problem into small parts
- Distribute to volunteers
- Solvable within minutes
- No training required

Experiment

- Online survey (~400 participants)
- Choose best building shape
- Vary number of buildings per task
- Measure time and accuracy

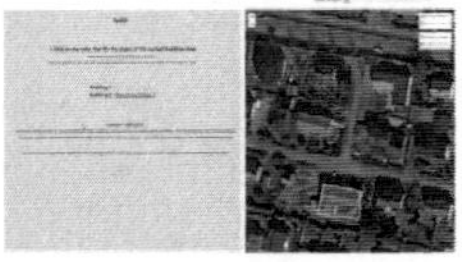

Results

Experience

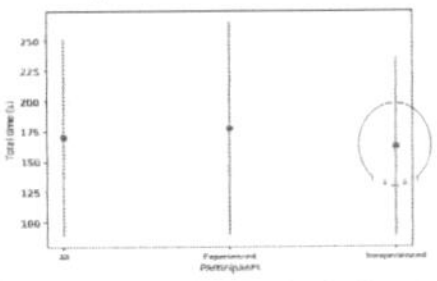

Inexperienced → faster

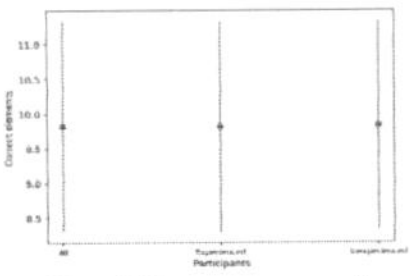

No difference on # correct

Buildings per task

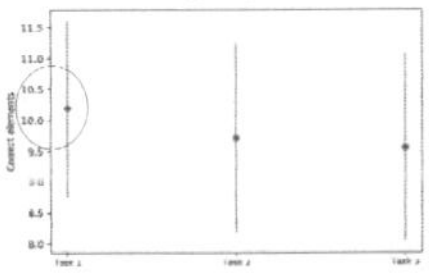

Fewer → more correct

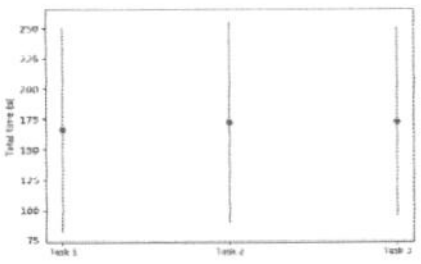

Time: no difference

Conclusion

- Micro-tasking is a feasible solution for OSM data imports
- Suitable for inexperienced participants
- Keep number of elements per task at minimum

More information at:
http://docs.atlefren.net/osm

About Us

Anne Sofie Strand Erichsen

MSc. Candidate
anneri@norkart.no

Atle Frenvik Sveen

Ph.d. Candidate
atlsve@norkart.no
@atlefren

Submitted to FOSS4G Boston: 03.06.2017

NORKART

NTNU
Norwegian University of Science and Technology

GIS Investigation of Crime Prediction with an Operationalized Tweet Corpus

Anthony J. Corso[a,*], Abdulkareem Alsudais[b]

[a] *California Baptist University*

[b] *Claremont Graduate University*

Abstract: Social media as the de facto communication channel is being used to disseminate one's diurnal self-revelations. This profound discovery often contains double-talk, peculiar insights, or contextual information about real-world events. Natural language processing is regularly used to uncover both obvious and latent knowledge claims within disclosures published amid the complex environment. For example, a perpetrator with first-hand knowledge of their criminal incident uses social media to post critical information about it. A geographic information system (GIS) is capable of large-scale point data analysis and possesses methods that enable dataset processing, evaluation, and automatic spatial visualization. Such an artifactfused with traditional environmental criminology theory and social mediaerects guidelines, tools, and models for substantive construction and evaluation of GIS crime analysis solutions. Provided the social media stream is timely and correctly processed, corrective action can be taken. The construction of a natural language processing social media annotation pipe identifies latent indicators extracted from a social media corpus and is an integral part of societal mishap prediction. Spatial visualizations and regression analyses were used to describe and evaluate project artifacts. As a result, a social media corpus was operationalized, and subsequently used as a proxy for a traditional environmental criminology risk layer in construction of a social media GIS crime analysis artifact. Using such multi-domain collaboration, the artifact was able to increase the predictive crime incident outcome with an overall R-squared increase of 21.94%. This result is the state-of-the-art; there are no other results to compare it to.

Poster Download: http://scholarworks.umass.edu/foss4g/vol17/iss1/21

*Corresponding author
Email address: acorso@calbaptist.edu (Anthony J. Corso)

GIS Investigation of Crime Prediction with an Operationalized Tweet Corpus

Anthony J. Corso, Ph.D.

California Baptist University

FOSS4G BOSTON 2017

Introduction

Social media (e.g., tweets) are the de facto communication channel to disseminate one's diurnal self-revelations. This profound phenomenon contains double-talk, peculiar insight, and contextual data or information about real-world events. Amid such complex and personal expose, natural language processing (NLP) techniques uncover both obvious and latent knowledge claims published within.

A geographical information system is capable of large-scale data analysis and possesses methods that enable dataset processing, evaluation, and spatial visualization. When fused with traditional research theory—such an artifact defines guidelines, algorithms, and models for substantive and predictive investigation.

Objectives

Despite a tweet's sparse content, NLP makes their use in a predictive GIS artifact feasible. For example, subsequent to processing, useful tweets are able to:

- **Predict the validity of a real-world event only recorded by observation of social media eyewitness; or**
- **Predict real-time trends by amalgamating social media with traditional social behavior variables.**

Thus, inquiry explores GIS outcomes when consuming "useful" or "not useful" tweets as identified via NLP techniques. In addition, a research framework illustrates social media being coalesced with other behavior variables and subsequently used as a social behavior GIS proxy layer.

Research Framework

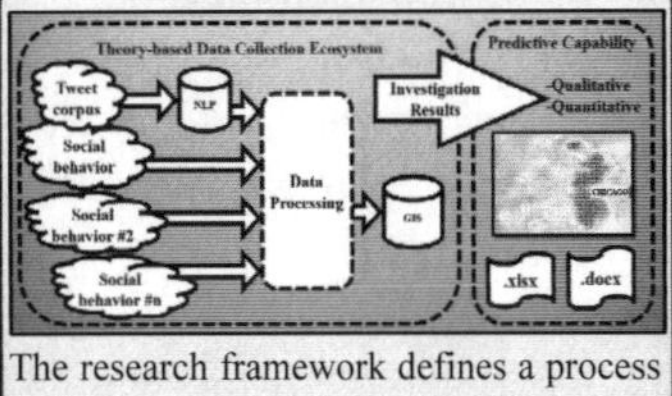

The research framework defines a process and exposes opportunity to fill the gap between sparse text social media and its representation of real-world events by examining meaningful tweet content and purging useless structures. That is, some tweets are so sparse they cannot represent the real-world context in which they exist; hence, a "Not Useful" tweet (illustrated in the table below). However, some tweets

Useful	Not Useful
I'm at Old Navy in Chicago IL https://t.co/lczpu9NLF	ballloooonnsss ??? http://t.co/mjhuKyH7DM
My Phone Die So Fast	WAYYOHANDSIDETOOSI
David Bowie is... my favorite! @ David wie is At Mca Chicago	Funniest
I aint gta stunt on nobody...trust me Yo LoL!	#CuppyCoffee!!!!!!!???
I can't wait to see lora tomorrow	@_lorShane ????

are "Useful" but require extra processing.

Tweets & Natural Language Processing

Operationalizing "useful" or "not useful" tweets was accomplished via the General Architecture for Text Engineering[1] (GATE) NLP suite of tools. The NLP pipeline built, as noted below, had three functions.

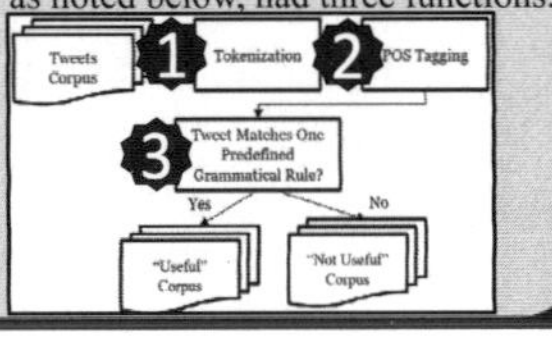

Therefore, it tokenized, part-of-speech tagged, and applied custom grammar rules to each tweet. A custom GATE NLP application (below) executed the pipeline.

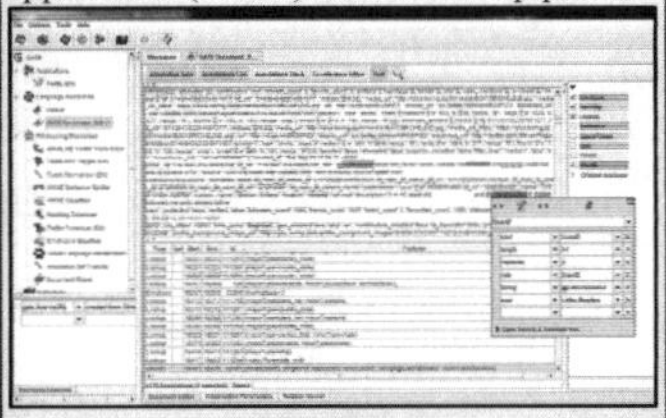

GIS Analysis

Association between features of a tweet, e.g., acronym use and its grammatical structure, and its potential usefulness were operationalized via NLP preprocessing. GIS capability examined both quantifiable and meaningful qualitative results; each are required in data analysis, information dissemination, and predictive artifacts.

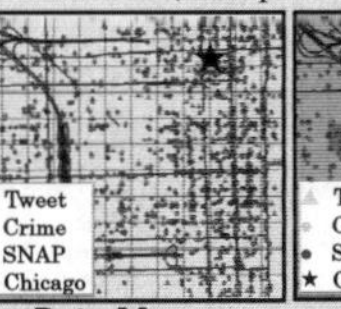

Data Map

Solution Map

The maps represent the area of downtown Chicago with a fishnet spacing of 750 feet. Tweets[2], crime[3] and SNAP[4] locations are the variables displayed. The Data Map is a visualization of the data. The Solution Map represents the results of a GIS grouping analysis tool used for exploratory variable analysis; the attributes are combined and cell shading represents latent structures.

Discussion and Conclusion

With a novel NLP pipeline tweets were processed and used to measure the change in performance of an ArcGIS[5] 10.4.1 artifact. A 1,000 tweet sample was hand tagged and compared to a baseline model, and to an innovative social media grammar applied by a rule-based social media NLP pipeline. GIS evaluation tools answer the question, prior to content analysis of a tweet, does a method exist to support identifying a tweet as "useful" for subsequent GIS processing? Indeed, "useful" tweet identification via NLP returned precision of 0.9256, recall of 0.6590, and F-measure of 0.7699; consequently, exploratory GIS processing of a social media variable increased 0.2194 over baseline.

Predictive capability potential of a GIS artifact implementing social media's latent behavior attributes is vast. Yes, preliminary results are encouraging but future research is important and needs to identify its value.

References

Alonso, O., Marshall, C. C., & Najork, M. (2013). Are some tweets more interesting than others?# hardquestion. Paper presented at the Proceedings of the Symposium on Human-Computer Interaction and Information Retrieval.

André, P., Bernstein, M., & Luther, K. (2012). Who gives a tweet?: evaluating microblog content value. Paper presented at the Proceedings of the ACM 2012 conference on Computer Supported Cooperative Work.

Bendler, J., Ratku, A., & Neumann, D. (2014). Crime Mapping through Geo-Spatial Social Media Activity.

[1] Bontcheva, K., Derczynski, L., Funk, A., Greenwood, M. A., Maynard, D., & Aswani, N. (2013). TwitIE: An Open-Source Information Extraction Pipeline for Microblog Text. Paper presented at the RANLP.

Bramsen, P., Escobar-Molano, M., Patel, A., & Alonso, R. (2011). Extracting social power relationships from natural language. Paper presented at the Proceedings of the 49th Annual Meeting of the Association for Computational Linguistics: Human Language Technologies.

Caplan, J. M., Kennedy, L. W., & Miller, J. (2011). Risk Terrain Modeling: Brokering Criminological Theory and GIS Methods for Crime Forecasting. Justice Quarterly, 28(2), 360-381. doi: 10.1080/07418825.2010.486037

[3] City of Chicago. (2014). Retrieved December 31, 2014, from https://data.cityofchicago.org/

Corso, A., Alsudais, K., & Hilton, B. (2016). Big Social Data and GIS: Visualize Predictive Crime. AMCIS Conference 2016 Proceedings.

Corso, A. J., & Alsudais, A. (2015). GIS, Big Data, and a Tweet Corpus Operationalized via Natural Language Processing. AMCIS Conference 2015 Proceedings.

Drawve, G. (2014). A Metric Comparison of Predictive Hot Spot Techniques and RTM. Justice Quarterly, 1-29. doi: 10.1080/07418825.2014.904393

[5] ESRI 2011. ArcGIS Desktop: Release 10. Redlands, CA: Environmental Systems Research Institute.

Frantzi, K., Ananiadou, S., & Mima, H. (2000). Automatic recognition of multi-word terms: the C-value/NC-value method. International Journal on Digital Libraries, 3(2), 115-130.

Gerber, M. S. (2014). Predicting crime using Twitter and kernel density estimation. Decision Support Systems, 61, 115-125.

Hirst, G., & Feiguina, O. g. (2007). Bigrams of syntactic labels for authorship discrimination of short texts. Literary and Linguistic Computing, 22(4), 405-417.

Hiruta, S., Yonezawa, T., Jurmu, M., & Tokuda, H. (2012). Detection, Classification and Visualization of Place-triggered Geotagged Tweets.

Hurlock, J., & Wilson, M. L. (2011). Searching Twitter: Separating the Tweet from the Chaff. Paper presented at the ICWSM.

Jurafsky, D., & Martin, J. H. (2009). Speech and Language Processing.

Kennedy, L. W., Caplan, J. M., & Piza, E. (2011). Risk Clusters, Hotspots, and Spatial Intelligence: Risk Terrain Modeling as an Algorithm for Police Resource Allocation Strategies. Journal of Quantitative Criminology, 27(3), 339-362. doi: 10.1007/s10940-010-9126-2

Levy, G. (2011). Designing User Studies in Informatics: Springer Science & Business Media.

Phan, X.-H., Nguyen, L.-M., & Horiguchi, S. (2008). Learning to classify short and sparse text & web with hidden topics from large-scale data collections. Paper presented at the Proceedings of the 17th international conference on World Wide Web.

Piao, S., & Whittle, J. (2011). A Feasibility Study on Extracting Twitter Users' Interests Using NLP Tools for Serendipitous Connections. Paper presented at the Privacy, security, risk and trust (passat), 2011 ieee third international conference on and 2011 ieee third international conference on social computing (socialcom).

Quattrone, G., Proserpio, D., Quercia, D., Capra, L., & Musolesi, M. (2016). Who Benefits from the "Sharing" Economy of Airbnb? Paper presented at the Proceedings of the 25th International Conference on World Wide Web, Montréal, Québec, Canada.

Sriram, B., Fuhry, D., Demir, E., Ferhatosmanoglu, H., & Demirbas, M. (2010). Short text classification in twitter to improve information filtering. Paper presented at the Proceedings of the 33rd international ACM SIGIR conference on Research and development in information retrieval, Geneva, Switzerland.

Stefanidis, A., Crooks, A., & Radzikowski, J. (2013). Harvesting ambient geospatial information from social media feeds. GeoJournal, 78(2), 319-338. doi: 10.1007/s10708-011-9438-2

Torres-Moreno, J.-M. (2014). Three Statistical Summarizers at CLEF-INEX 2013 Tweet Contextualization Track. Paper presented at the CLEF (Working Notes).

[2] Twitter. (2014). Retrieved December 31, 2014, from http://www.twitter.com

[4] United States Department of Agriculture. (2014). Retrieved December 31, 2014, from http://www.fns.usda.gov/snap/retailerlocator

Zingla, M. A., Chiraz, L., Slimani, Y., & Berrut, C. (2015). Statistical and Semantic Approaches for Tweet Contextualization. Procedia Computer Science, 60, 498-507.

Zubiaga, A., & Ji, H. (2014). Tweet, but verify: epistemic study of information verification on Twitter. [journal article]. Social Network Analysis and Mining, 4(1), 1-12.

Contact

Dr. Corso holds Ph.D. in Information Systems and Technology from Claremont Graduate University. Since 2007, he is an Associate Professor in the Gordon and Jill Bourns College of Engineering at California Baptist University.

E-mail: acorso@calbaptist.edu

Benchmark: using sensors to study public space

Mario A. Giampieri[a,*], Sarah Williams[a], Wenfei Xu, Hayrettin Gunc, Christine M. Langston[a], Dennis C. Harvey[a]

[a] *Massachusetts Institute of Technology*

Abstract: Efforts have been made throughout history to measure how people use public space. This research seeks to integrate a range of sensor technologies to automate analysis of pedestrian usage of public space. A range of environmental sensors, image recognition utilities, and open-source software are combined to create a system to measure in detail how people use public space, with the intention of serving as a tool for creating better public spaces in the future. This framework is part of a broader effort to offer organizations and individuals methods and data to inform place-making interventions at multiple scales in conjunction with the Gehl Institute and Better Block Foundation.

Poster Download: http://scholarworks.umass.edu/foss4g/vol17/iss1/22

*Corresponding author
Email address: marioag@mit.edu (Mario A. Giampieri)

Benchmark: using sensors to study public space

Mario Giampieri, Sarah Williams, Wenfei Xu, Hayrettin Gunc, Christine Langston, Dennis Harvey MIT Civic Data Design Lab

Introduction

Efforts have been made throughout history to measure how people use public space. This research seeks to integrate a range of sensor technologies to automate analysis of pedestrian usage of public space. A range of environmental sensors, image recognition utilities, and open-source software are combined to create a system to measure in detail how people use public space, with the intention of serving as a tool for creating better public spaces in the future.

Sensing was divided into local and site-wide categories. Mobile sensing was achieved through using force sensors, magnets, GPS, and photosensors. Fixed sensing was achieved through the use of a camera and image recognition packages. This poster details the fixed sensing.

mobile sensing

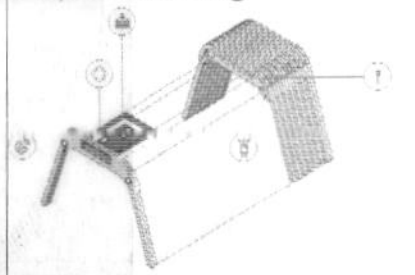

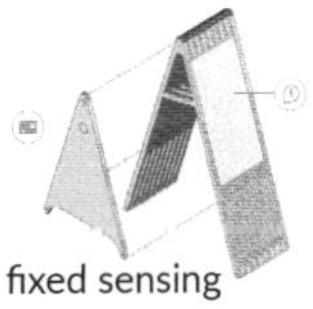

fixed sensing

Methods

A number of fixed information boards were placed throughout the site, each outfitted with a GoPro camera attached to a Raspberry Pi model-B. OpenCV 3.2.0 (opencv.org) was used to identify human faces from video feeds and estimate the location of the persons in the video feed relative to the camera.

Example scenario showing pedestrian detection and position relative to camera in feet

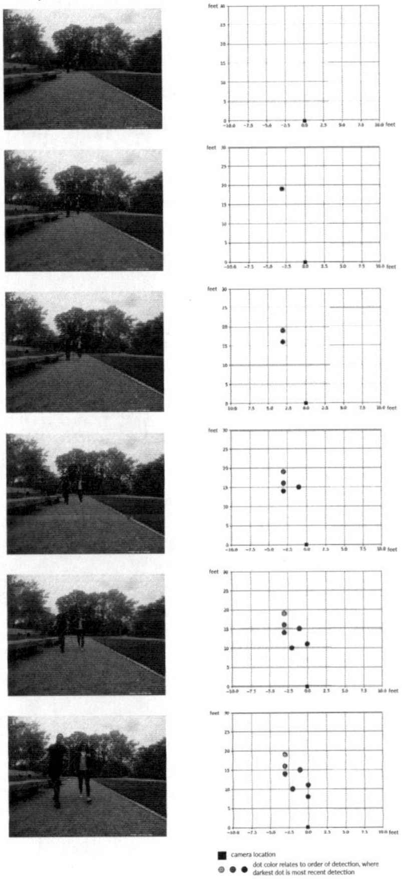

Results

Positive pedestrian detection was achieved in about half of images collected in the two week trial period. The images and plots (created using Pandas Python library: pandas.pydata.org) to the left illustrate a typical scenario where pedestrians are detected within 30 feet of the camera. The boxes drawn around faces denote positive detection, and the lines represent the geometries used to calculate relative distance from the camera.

Locations are reported relative to the camera in a two-dimensional plane. Reported pedestrian locations were within one foot of measured distances in test scenarios, suggesting promise for this method once higher detection rates are attained through iteratively training the classifier. Translating from relative to geographic coordinates is straightforward and is an area of future development.

Discussion

Our application focuses the potential of computer vision software toward understanding how people use public space in concert with other local sensors and place-making interventions. This combination will enable urban planners and designers to study public spaces effectively and at a wider scale than before. This solution is low-cost, and relies upon easily available materials and open-source code[1], suggesting that organizations interested in studying public space (and arguing for place-making interventions) can do so more easily than by using traditional methods. An ideal use case would be analyzing an area before and during a temporary intervention to demonstrate the benefits of peace-making interventions and to help make such interventions more permanent.

1. OpenCV team, 2016. Open Source Computer Vision Library. www.opencv.org.

Kadaster Data Platform - Overview Archicture

Erwin Folmer[a,*], Wouter Beek[b]

[a] *Kadaster, University of Twente*

[b] *Kadaster, VU University Amsterdam, Triply*

Abstract: The Dutch Cadastre is publishing its geospatial data assets as Linked Open Data through the Kadaster Data Platform (KDP). The KDP supports the following three Linked Data browsing paradigms: (1) graph navigation, (2) hierarchical browsing, and (3) faceted browsing. Graph navigation uses the graph-shape of the RDF datamodel, to display concepts and instances as nodes, and properties between them as edges between those nodes. Graph navigation works well for explorative browsing. For graph navigation the KDP uses LODView (http://lodview.it), an existing OSS. Hierarchical browsing uses the tree structure of the concept hierarchy in order to display the various classes and properties that are in the data. A hierarchical browser gives the user a quick overview of the main classes and properties that are in a dataset. Hierarchical browsing works well for gaining an understanding of a concept schema. For this the OSS Linked Data Theatre is being developed (https://github.com/architolk/Linked-Data-Theatre). Faceted browsing is an advanced data browsing technique that turns the properties that appear in a dataset into widgets that can be set by the user in order to formulate a conjunctive set of filters over the set of instances that is described in the data. Faceted browsing works well for searching a specific instance. Triply has built FacetCheck, faceted browser that configures itself. For querying the KDP implements YASGUI: an integrated web service based on YASQE (a query editor that provides direct feedback to the user) and YASR (a versatile query results visualizer). Visit https://data.pdok.nl/yasgui to query billions of geospatial objects through the YASGUI query editor and result set visualizer!

Poster Download: http://scholarworks.umass.edu/foss4g/vol17/iss1/23

*Corresponding author
Email address: `erwin.folmer@kadaster.nl` (Erwin Folmer)

Kadaster Data Platform

Overview Architecture

Erwin Folmer | Kadaster | University of Twente | erwin.folmer@kadaster.nl **Wouter Beek** | Kadaster | VU University Amsterdam | Triply | wouter@triply.cc

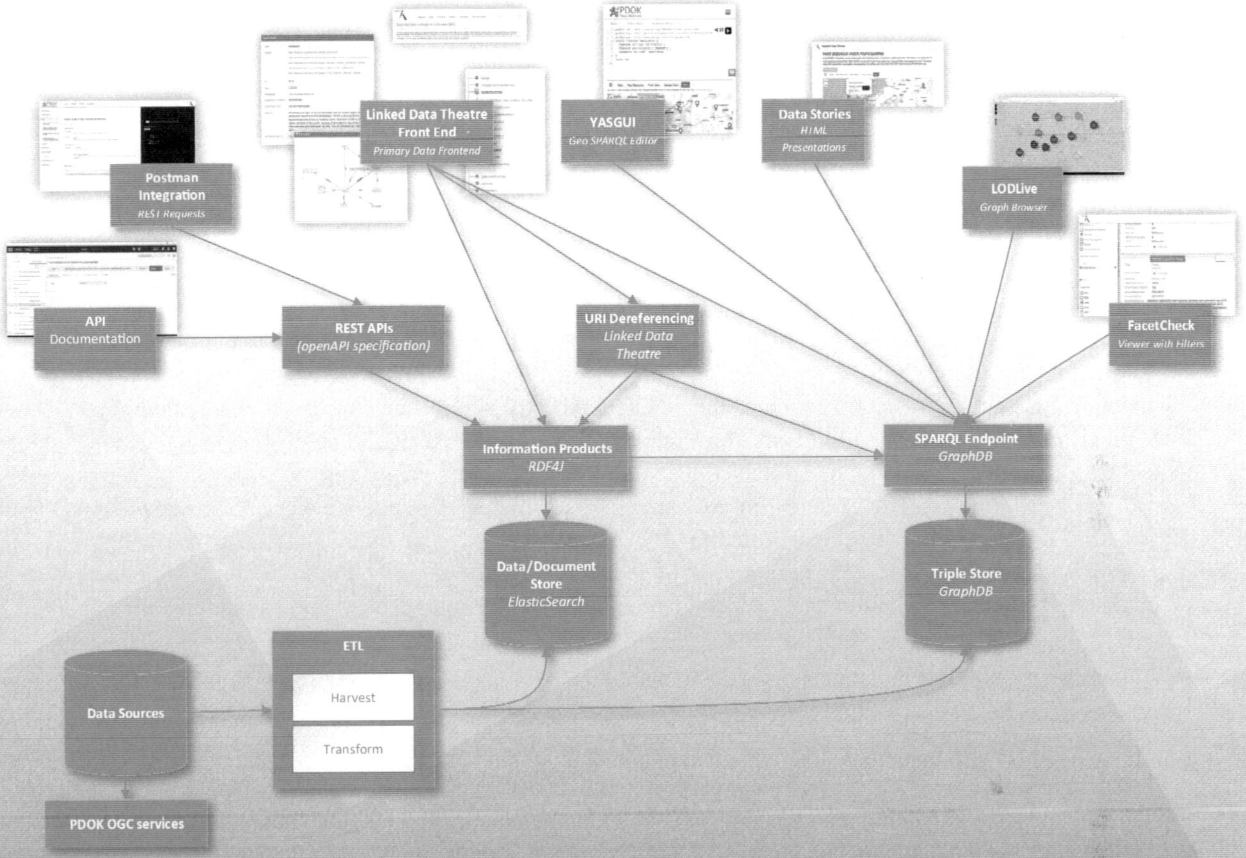

Dutch Cadastre is developing an innovative data platform for publishing its geospatial data as linked data.

Results so far:

- Platform released at data.pdok.nl
- 3 formal key registers of the Dutch goverment published
- Billions of RDF triples
- Hundreds of millions of geospatial objects
- Actual data (by having daily updates)
- APIs available through openAPI specification, including documentation and postman integration
- Open SPARQL endpoint
- Forum for direct interaction with users

www.kadaster.nl/dataplatform of data.pdok.nl

kadaster PDOK Publieke Dienstverlening op de Kaart

Open Source and Open Learning

Ryan Beare[a,*], Vidhi Goti[a], Seth Trojacek[a]

[a] *Texas Am University Commerce GIS Club*

Abstract: We are the Geospatial Information Systems Club from Texas AM University - Commerce in Commerce, TX. Over the past year, we have cultivated an incredible community of researchers, data analysts, programmers, and writers to make our club one of the top academic clubs on our campus. Our first project even won a first place award. How did we do it? We used Foss4G software. As beginners, we came together as a community of close friends, learned QGIS and undertook a project that highlighted and celebrated the rich 125 year history of our university. Together we continue to use QGIS to study such topics as gentrification, education scarcity, cattle migration, movement of our alumni over time, and many others. We take a different educational approach to learning. Rather than setting tight deadlines for large projects, we come together, we collaborate to solve problems. In addition to our use of QGIS, we have developed our own open source software, and we only use open source software when completing our projects. We want to show how learning GIS looks outside of the classroom, and how beneficial the community of problem solvers we've created has been to our university.

Poster Download: http://scholarworks.umass.edu/foss4g/vol17/iss1/24

*Corresponding author
Email address: `rnbeare@gmail.com` (Ryan Beare)

TEXAS A&M UNIVERSITY COMMERCE

OPEN SOURCE AND OPEN LEARNING

TEXAS A&M UNIVERSITY- COMMERCE

GEOSPATIAL INFORMATION SYSTEM AND DATAVISUALIZATION CLUB

FOSS4G BOSTON 2017

AUGUST 16TH-18TH
BOSTON, MA.

BACKGROUND

We are the Texas A&M University-Commerce Geospatial Information Systems Club, located in Commerce, Texas. Academic use of geospatial information systems is a common thing in American universities. This club takes the academic approach outside of the classroom, using open source software and technology. We decided to start this club as a university-wide opportunity to come together and recognize our heritage as one of the State's oldest learning institutions. We have done research spanning all of Hunt county, assisting the city of Commerce in research, as well as, parts of the greater Dallas area. Most recently, we have completed a project that placed all of our alumnus on a map of the entire world, which we completed using only open source software. We accomplished this by taking a different route than the traditional method of education. We applied Geospatial analysis to locate the alumni using QGIS. We claim that it may be the better way to learn Geospatial Information.

OUR APPROACH

Being one of the oldest learning institutions in Texas has allowed us to research a very large amount of data, which might not be possible for other schools who may not have been in existence. Since, Texas A&M Commerce has such extensive data on their alumni, we can provide a more holistic understanding of where our students have come from. Open source software, such as QGIS helps make research like this possible, because of how accessible it is. Our aim from the beginning has been to be as accessible as possible so that we can all leave our mark through collaboration

OUR RESULTS

The results we experience from practicing an open source attitude towards learning for two semesters have been beyond impressive. Overall, we are the fastest growing academic and research based club on our university campus. We expect the growth to continue into fall 2017. Through collaboration, participation, critical thinking and problem solving we have been able to recently complete the largest project of mapping over 164,000 of our school's alumni from all across the world. Our coding team has produced our own website, which we plan to launch in July, as well as two mobile games that will serve as fundraising for our club's future expenses.

OUR MEMBERS

Today, our club is almost 20 members strong. Our members come from as near as Commerce, Texas to as far away as India. This helps us create a multicultural experience for everyone involved. Currently, no one in our club has previous extensive background in GIS.

Seth Trojacek – Seth is a certified Early Childhood Educator, and a current Computer Science Graduate student at Texas A&M University- Commerce, specializing in Artificial Intelligence. He also looks into the Education theory behind GIS and works on incorporating teaching principles to the groups projects.

Vidhi Goti – Vidhi is pursuing her Master's in Business Analytics at Texas A&M University-Commerce. Prior to joining grad school, she worked at CRISIL- An S&P Global Company under structured finance domain where she worked on multiple technologies like Informatica PowerCentre, and SQL.

ALUMNI PROJECT

Below is the result of our first collaborative effort in which we came together and plotted Texas A&M University Commerce Alumni data. This project received a first place award at our school's Lion's Innovation Showcase competition.

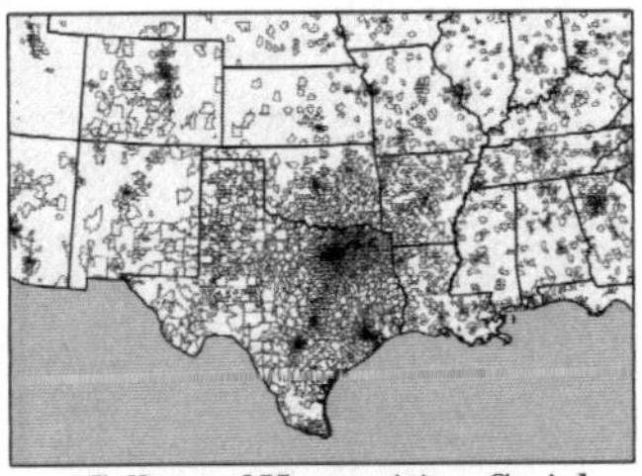

College of Humanities, Social Sciences and Arts

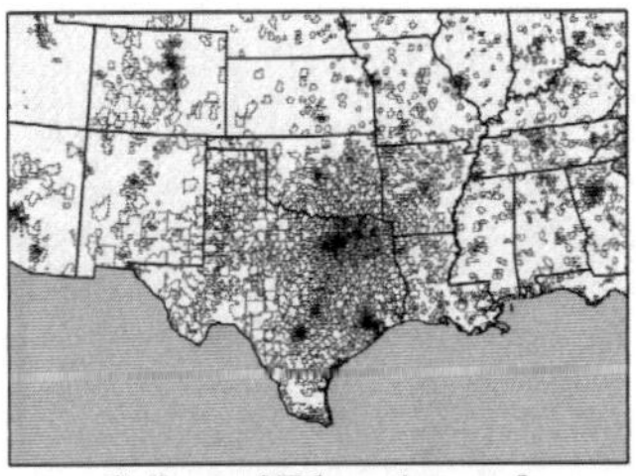

College of Education and Human Services

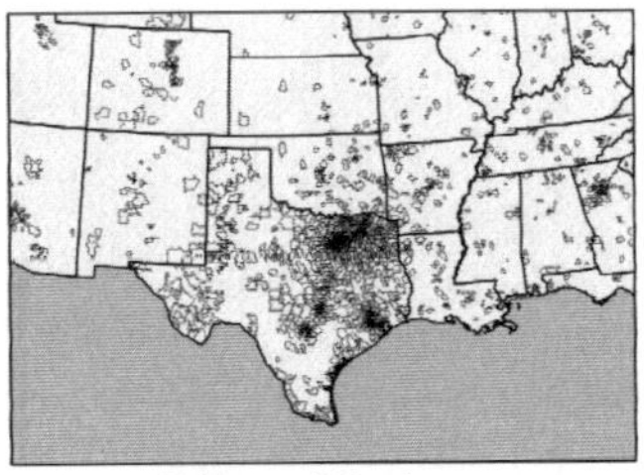

College of Business

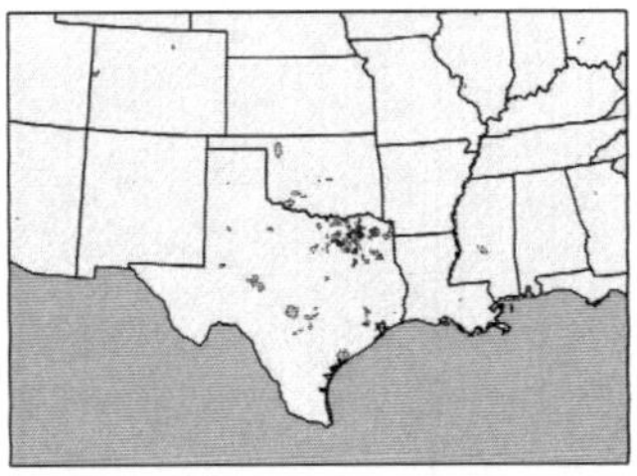

College of Science and Engineering

Optimum Object Analysis Of Islands Activities On South China Sea By DNB On VIIRS

Ichio Asanuma[a,*], Takashi Yamaguchi[a], Jong-geol Park[a], Keneth J. Mackin[a]

[a] *Tokyo University of Information Sciences*

Abstract: A data processing and analyzing system was designed and made operational with free software to process the big data over the South China Sea (SCS), which are provided by the Visible Infrared Imaging Radiometer Suite (VIIRS) on the Suomi National Polar-orbiting Partnership (Suomi-NPP) to monitor the light distributions in the night. The VIIRS data are processed from the raw data (level-0) to geophysical data (level-3) using the International Polar Orbiter Processing Package (IPOPP), which are freely available from NASA. The Day Night Band (DNB) of VIIRS is extracted from the level-3 data and is geospatially projected for our region of interest (ROI) on the SCS. For those ingested data, an optimum object analysis in geographic domain was proposed to estimate the reclamation activities on coral reefs in the SCS using GDAL. A pixel base analysis of DNB data is possible to estimate the island activities independently among dredgers, support vessels, or buildings on coral reefs, but not appropriate to analyze the activities for as an integrated system or for the changes in the ROI. Although there is a difficulty to determine the scale of objects on the analysis of DNB data for the reclamation activities, the optimum object scale was empirically determined for different size of coral reefs and was applied for this study. The optimum object analysis determines the reclamation activities with including lights not only from buildings but also dredgers and supporting vessels around coral reefs.

Poster Download: http://scholarworks.umass.edu/foss4g/vol17/iss1/25

*Corresponding author
Email address: `asanuma@rsch.tuis.ac.jp` (Ichio Asanuma)

Optimum Object Analysis Of Islands Activities On South China Sea By Day Night Band (DNB) On Visible Infrared Imaging Radiometer Suite (VIIRS)

Ichio Asanuma, Takashi Yamaguchi, Jong-geol Park, and Keneth J. Mackin
Tokyo University of Information Sciences

Objectives

Objectives of this study are to detect the island activities and to monitor a change of fishing activities on the Spratly Islands in the South China Sea, Using the Day Night Band (DNB) [1,2] of the Visible Infrared Imaging Radiometer Suite (VIIRS) on the Suomi National Polar-orbiting Partnership (Suomi-NPP).

Region Of Interest (ROI)

The Spratly Islands are one disputed area in the world among nations faced into the South China Sea. Recently, China reclaimed several coral reefs, which escalated the conflictions among them.

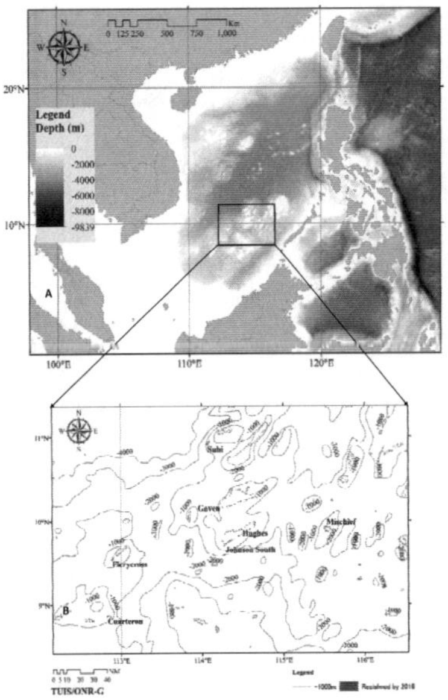

Fig.1 Region of interest (ROI) of Spratly Islands in the South China Sea, 7 reefs were reclaimed by China.

Fig.2 Tokyo University of Information Sciences (TUIS) operates three receiving stations in Japan. The station on the Miyakojima Island covers the Northern half of South China Sea.

Method

Raw data to level-2 data

International Polar Orbiter Processing Package (IPOPP), FOSS provided by NASA, ingests the raw data of VIIRS, received by the ground stations, and process to level-2 (geophysical data) through level-0 and -1 [3].

Level-2 to level-3 data

The DNB, 500 to 900 nm, and M12, 3.66 to 3.84 μm, are extracted from 22 multiple bands to identify lights from surface, and geospatially projected on to the map over the ROI using **Geospatial Data Abstraction Library (GDAL)**.

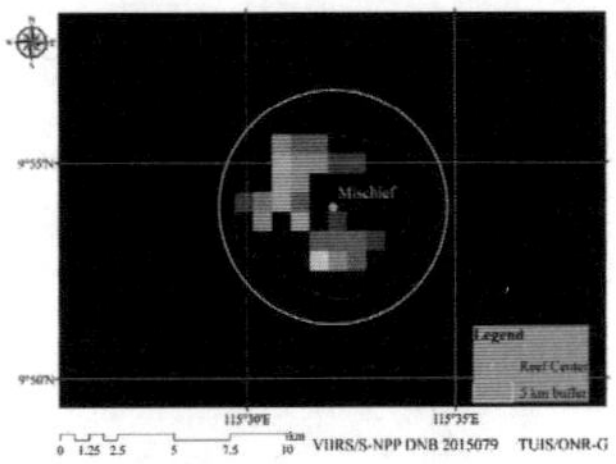

Fig.3 Lights distribution detected by DNB over Mischief Island with a buffer of 5 km radius from the center of island, observed on March 19, 2016.

The lights from the surface without the lunar illumination could be easily identified by DNB, but the lunar illumination makes it difficult when lights reflected by clouds exist.

Change of mean intensity within buffer

The mean intensity within the buffer was obtained by **QGIS** for the reclaimed islands by the Chinese government from Jan. 1st of 2014 to Dec. 31st, 2016. The buffer is defined to include lights on the islands and vessels around the island as an optimum size of buffer based on objects.

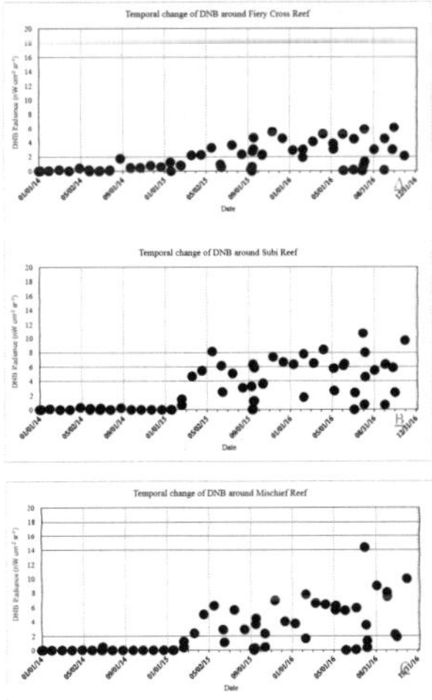

Fig.4 History of mean lights from the optimum size of buffer.

Results

A slightly increase in DNB mean intensity was observed at the beginning of dredging at Subi Reef and Mischief Reef. Similarly, a sharp reduction in DNB mean intensity was observed at the end of dredging at Subi Reef and Mischief Reef, but was that drop in intensity was not as clear at Fiery Cross Reef. It is estimated that the dredging period at Subi and Mischief Reefs were about 6 months each.

References

[1] Levin, N., Yishai Duke, 2012, High spatial resolution night-time light images for demographic and socio-economic studies, Remote Sens. of Env., 119, 1-10.

[2] Elvidge, C. D., M. Zhizhin, K. Baugh and F. Hsu, 2015, Automatic Boat Identification System for VIIRS Low Light Imaging Data, Remote Sens., 7(3), 3020-3036.

[3] Asanuma, I., T. Yamaguchi, J. Park, K. J. Mackin, J. Mittleman, 2016, Detection of fishing boats by the day night band (DNB) on VIIRS, Imaging Spectrometry XXI, Proc. of SPIE, 9976, 99760P.

Acknowledgement

This research is partly supported by the Office of Naval Research Global of the USA with the grant number of N62909-15-1-2074.

Supporting Trajectory UDF Queries and Indexes on PostGIS

Pyoung Woo Yang[a,*], Kwang Woo Nam[b]

[a] *Turbo Soft Inc.*

[b] *Kunsan National University*

Abstract: In this paper, we propose a system model for querying and indexing the GPS trajectory of moving objects on PostGIS/PostgreSQL. We developed moving object data types including MPoint(moving point), MDouble(moving double) for GPS trajectories. Also, various moving objects UDFs(user-defined functions) are implemented for moving objects queries. For efficient query processing, r-tree index is extended for trajectory, and pre-materialization techniques are proposed for fast UDF processing. Experimental results show that the pre-materialization techniques are about 1.2 times faster than nave query processing using r-tree index.

Poster Download: http://scholarworks.umass.edu/foss4g/vol17/iss1/26

*Corresponding author
Email address: `manner7979@gmail.com` (Pyoung Woo Yang)

Supporting Trajectory UDF Queries and Indexes on PostGIS

Pyoung Woo Yang* and Kwang Woo Nam**

*Turbo Soft Inc. and **Kunsan National University

kwnam@kunsan.ac.kr

Introduction

In this paper, we propose a system model for querying and indexing the GPS trajectory of moving objects on PostGIS/PostgreSQL.

We developed moving object data types including moving point, MDouble for GPS trajectories. Also, various moving objects UDFs are implemented for moving objects queries. For efficient query processing, r-tree index is extended for trajectory, and pre-materialization techniques are proposed for fast UDF processing.

Experimental results show that the pre-materialization techniques are about 1.2 times faster than naïve query processing using r-tree index.

Related Work

UDF(User Defined Function) approach have been used widely for developing advanced applications[1]. But, trajectory DBMS is not popular because of performance optimization [2].

Trajectories DBMSs

- DOMINO[3]
- HERMES[4]
- Trajectory Functions of PostGIS[5]

Trajectory UDF Queries by Examples

Creation of Trajectory Table

```
CREATE TABLE taxi
( taxi_id int, taxi_number char(20),
  taxi_model char(20), taxi_driver char(20) );

SELECT addtrajectorycolumn('public', 'taxi', 'traj', 4326, 'MOVINGPOINT', 2, 150);
```

Appending/Updating Trajectories

```
## Inserting Moving Objects
insert into taxi values(1, '57NU2001', 'Optima', 'hongkd7');
insert into taxi values(2, '57NU2002', 'SonataYF', 'hongkd7');

## Appending and Updating GPS Trjaectories

UPDATE taxi
SET traj = append(traj, 'MPOINT( 100 100 5000, 150 150
5001)') WHERE taxi_id = 1;

UPDATE taxi
SET traj = remove(traj, 'PERIOD( 5001, 5003)' )
WHERE taxi_id = 1;
```

Retrieving Trajectories

```
## Spatial and Temporal Slicing
SELECT [illegible]
FROM taxi
WHERE [illegible]

## Trajectory Predicates
SELECT [illegible]
FROM taxi
WHERE [illegible]

## Trajectory Functions
SELECT taxi_id, taxi_number, TJ_Traj(traj)
FROM taxi
WHERE TJ_Passes(traj, TJ_BOX(116.35, 39.93, 116.22, 40.14,
      PERIODS('2008-02-02 13:30:44', '2008-02-02 15:54:46')));

SELECT count(*)
FROM taxi
WHERE TJ_Inside(traj, TJ_BOX(116.35, 39.93, 116.22, 40.14,
      PERIODS('2008-02-02 13:30:44', '2008-02-02 15:54:46')));

SELECT taxi_id, taxi_number
FROM taxi
WHERE TJ_Cross(traj, taxi_id, TJ_BOX(116.35, 39.93, 116.22, 40.14,
      PERIODS('2008-02-02 13:30:44', '2008-02-02 15:54:46')));

## MDouble Functions

SELECT taxi_id, m_distance(traj, GeomFromText('Point( 50 50 )')),
      m_mindistance(traj, GeomFromText('Point( 50 50 )')),
      m_maxdistance(traj, GeomFromText('Point( 50 50 )'))
FROM taxi;

SELECT [illegible]
FROM taxi
WHERE [illegible]
      GeomFromText('Point( 50 50 )')) < 50;
```

Trajectory Data Model on PostGIS

Trajectories are split into trajectory segment tables. And, metadata for trajectory objects are stored in a trajectory_column table.

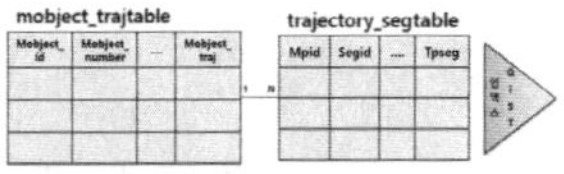

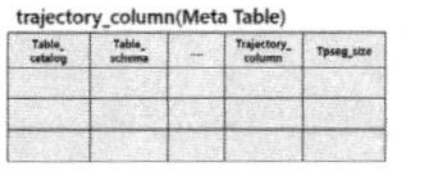

Performance Optimization by Query Materialization

Naïve Approach : No Optimization

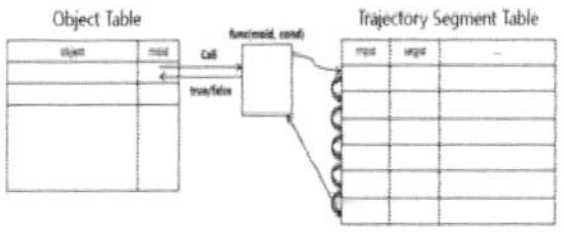

GiST Index Extension Approach for Trajectories

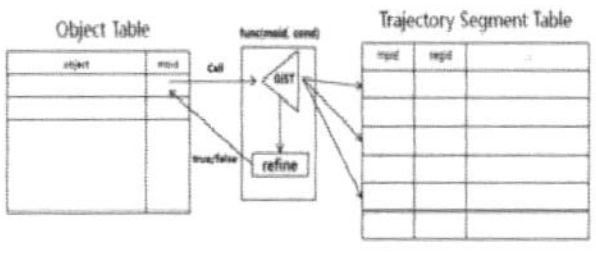

Query Materialization Approach

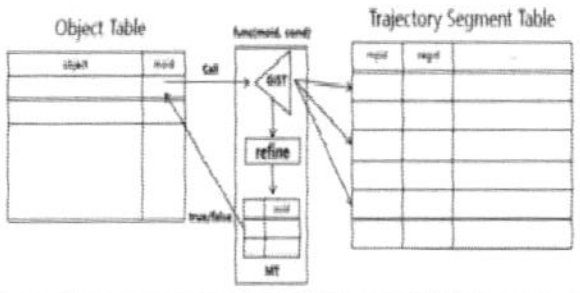

Performance Evaluation

- Time by number of trajectories

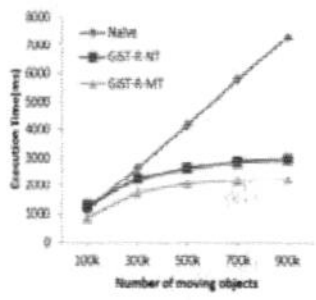

- Time by query window size

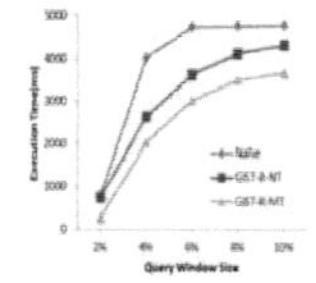

Conclusion and References

PostTrajectory : We developed a new system supporting trajectory queries on PostGIS using UDFs.

[1] C. Ordonez, and C. Garcia-Alvarado, "A data mining system based on SQL queries and UDFs for relational databases," in Proceedings of the 20th ACM Conference on Information and Knowledge Management, 2011, pp. 2521-2524.

[2] O. Wolfson, B. Xu, S. Chamberlain, and L. Jiang, "Moving objects databases: Issues and solutions," in Proceedings of the 10th International Conference on Scientific and Statistical Database Management, 1998, pp. 111–22.

[3] O. Wolfson, A.P. Sistla, B. Xu, S.J. Zhou, S. Chamberlain, "DOMINO: databases for moving objects tracking," in Proceedings of the SIGMOD International Conference on Management of Data, 1999, pp. 547–549.

[4] N. Pelekis, Y. Theodoridis, S. Vosinakis, T. Panayiotopoulos, "Hermes - A Framework for Location-Based Data Management," in Proceedings of EDBT, 2006, pp. 1130-1134.

[5] https://postgis.net/docs/reference.html#Temporal

ACKNOWLEDGEMENT

This research was supported by a Grant (14NSIP-B080144-01) from National Land Space Information Research Program funded by Ministry of Land, Infrastructure and Transport of Korean government.

PostTrajectory Project
http://github.com/awarematics/posttrajectory

The Development of Japanese City's Future Simulation System: My City Forecast

Toshikazu Seto[a,*], Hiroshi Omata[b], Yuki Fukushima[a], Yoko Hasegawa[b], Midori Maeda[b], Yoshihide Sekimoto[b]

[a] *Center for Spatial Information Science, the University of Tokyo*

[b] *Institute of Industrial Science, the University of Tokyo*

Abstract: In recent years, the optimization of urban management due to the rapid population decline has been one of the major issues in Japan. Future population estimation and related statistical information, such as the location information of the public facility is now available to open. However, open data utilization in the urban planning field is not advanced in comparison with other countries. We constructed the Web system using FOSS4G that citizens can be the future image of the city to operate on their own. It used mainly below FOSS4G tools; OpenLayers, PostGIS, Pgrouting and Geocolor. The collected data calculated in advance, and the default mode for displaying a simulation result in the 500m mesh unit. In addition to there is a customized mode to be re-calculated using the geodata that citizen is uploading open geospatial data. In the customization mode, to choose the residence induction region arbitrarily, it is possible to more detailed simulation by inputting the future urban areas. We also held a workshop for Japanese citizens the variety of stakeholders, it was also able to get feedback on the functional requirements.For the ordinary citizens are not familiar with GIS, to be able to display an easy-to-understand the future of the region have been evaluated. However, also revealed that the operation of the customized version feels as difficult to use, such as difficult legend of adjustment.

Poster Download: http://scholarworks.umass.edu/foss4g/vol17/iss1/27

*Corresponding author
Email address: `tosseto@csis.u-tokyo.ac.jp` (Toshikazu Seto)

Center for Spatial Information Science, The University of Tokyo · 東京大学 空間情報科学研究センター
OSGeo.JP · FOSS4G BOSTON 2017

The Development of Japanese City's Future Simulation System: MyCityForecast (https://mycityforecast.net/)

Toshikazu Seto[1], Hiroshi Omata[2], Yuki Fukushima[1], Yoko Hasegawa[2], Midori Maeda[2] and Yoshihide Sekimoto[2]

1: Center for Spatial Information Science, the University of Tokyo, Email: tosseto@csis.u-tokyo.ac.jp

2: Institute of Industrial Science, the University of Tokyo

Introduction

In recent years, the optimization of urban management due to the rapid population decline has been one of the major issues in Japan. Future population estimation and related statistical information, such as the location information of the public facility is now available to open.

However, open data utilization in the urban planning field is not advanced in comparison with other countries and big gap between planner and citizen for urban planning in Japanese local government.

Image of compact city (From Ministry of Land, Infrastructure, Transport and Tourism)

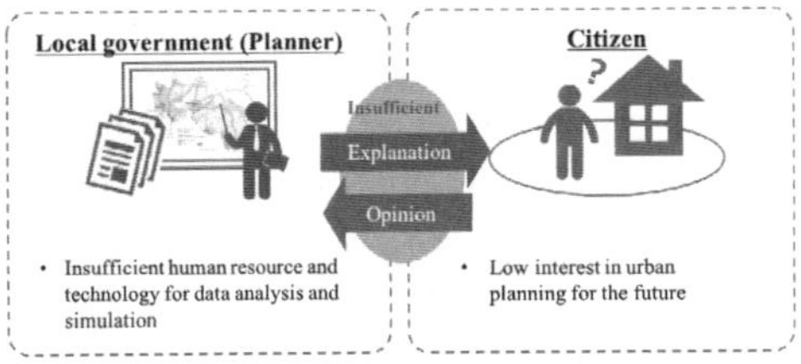

We developed "My City Forecast" for urban design communication of our own cities

MyCityForecast (MCF)

We constructed the Web system using FOSS4G that citizens can be the future image of the city to operate on their own. The collected data calculated in advance, and the default mode for displaying a simulation result in the 500m mesh unit.

* We are developed and opened 1,670 local governments; 97% of all Japanese cities.

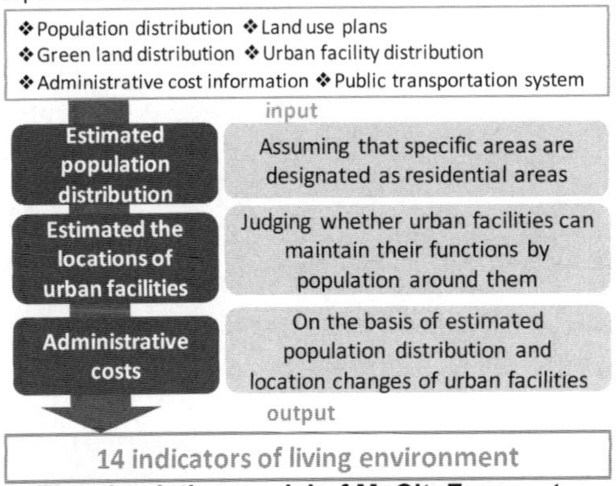

The simulation model of MyCityForecast

Main Framework	Library (JavaScript)
Php 5.5	OpenLayers.js
Mono 5.0.1	D3.js
PostgresSQL 9.3	Geocolor.js
- Postgis 2.0	Geo_tools.js
- pgrouting 2.0	Codemirror.js
gdal/ogr 2.2.0	Jquery (Jqplot)

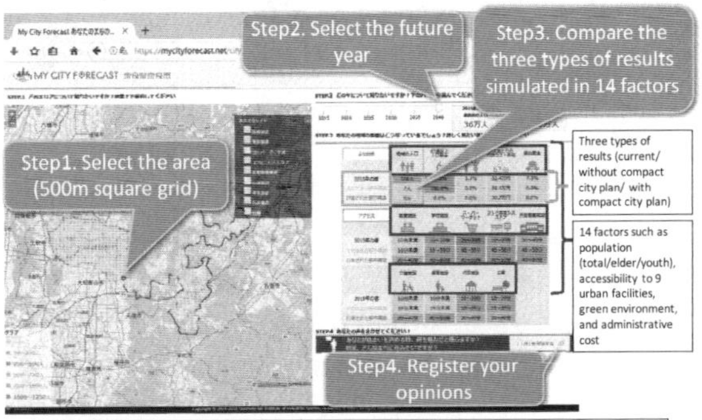

2015 value	Current status
Business As Usual Ver.	Case without any plans (BAU)
Compactized Ver.	Case with concept of compact city

The basic web interface of MyCityForecast

MCF Customize Version

"customization" allows the user to tailor the system to local (geographic) areas and conditions. In particular, the "residential advancement district" has different ideas in various peoples situation, and re-simulation is necessary.

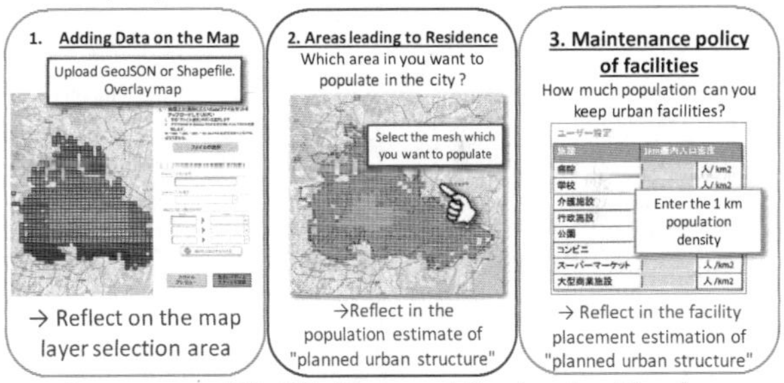

Summary of MyCityForecast Customize Version

MCF Workshop with Citizens

We held a workshop with the administration - citizens in three areas this year (Okayama, Shimane and Toyama). In particular, the simulation customizing the residential guidance area creates an important discussion for future planning. It is expected to be used in the policy of aging, disaster prevention, child rearing.

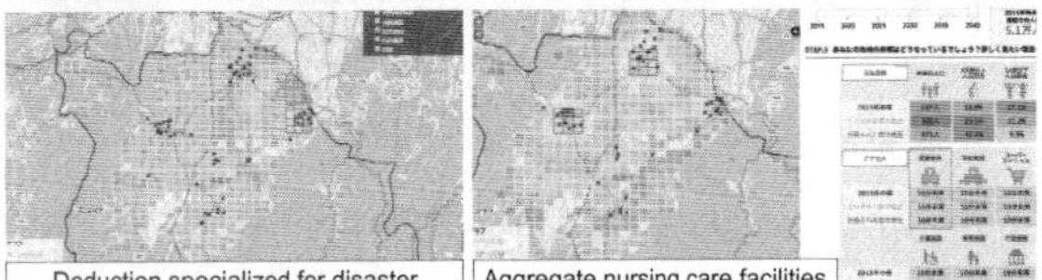
Deduction specialized for disaster countermeasures · Aggregate nursing care facilities and medical facilities

Using OSGeo solutions for local development systems implementation. The experience for the Northern Region of Costa Rica

Lpez-Villegas Oscar[a,*], Vquez-Acua Oscar[a], Vquez-Acua Leonardo[a]

[a]*Costa Rica Institute of Technology*

Abstract: Although some general definitions classify Spatial Data Infrastructures (SDI) as technological standards, institutional and even political agreements, which allow the discovery and use of geospatial information by users for different purposes [Kuhn 2005], computationally this platforms are valuable data repositories that should reach people efficiently and effectively for analysis and decision making on issues of collective interest. Costa Rica has several SDIs experiences at national level (SNIT - http://www.snitcr.go.cr), regional level (IDEHN - http://www.idehn.tec.ac.cr) or local/cantonal level (IDESCA - http://idesca.cr). Those infrastructures can facilitate access between geospatial information managers and their consumers through the implementation of particular software applications. The following work shows the experience of developing applications supported by open source software solutions (mainly OSGeo's software), consuming local SDIs data in accordance with the realities of the region and geospatial data producers.

Poster Download: http://scholarworks.umass.edu/foss4g/vol17/iss1/28

*Corresponding author
Email address: `olopez@tec.ac.cr` (Lpez-Villegas Oscar)

Using OSGeo solutions for local development systems implementation

The experience for the Northern Region of Costa Rica

O. Víquez, L. Víquez, O. López

Computer Science Department, Costa Rica Institute of Technology, San Carlos Campus

ABSTRACT

Although some general definitions classify Spatial Data Infrastructures (SDI) as technological standards, institutional and even political agreements, which allow the discovery and use of geospatial information by users for different purposes [Kuhn 2005], computationally these platforms are valuable data repositories that must reach people efficiently and effectively for analysis and timely decision-making on issues of collective interest.

Costa Rica has several SDIs experiences, the following work shows the experience for developing applications supported by OSGeo's software, consuming several local SDIs data in accordance with the realities of the region and geospatial data producers.

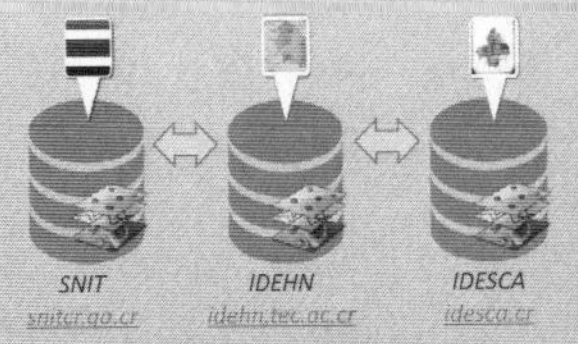

OBJECTIVES

Our main objective was to contribute to the improvement of SDIs geospatial data access through the implementation of software applications using open source standards and software technologies.

ACCESSING SDI GEOSPATIAL DATA

Geoservices information sources can range from files to a collection of different geographic databases and even information from other types of remote sensors [Arnhardt, 2007]. Geoservices make their sources transparent to the user; however, they could be limited to some information geoprocessing capabilities that could be performed directly from the database.

Several experiences were developed from using WMS and WFS protocols to manipulate geographic data layers to direct database query connections, allowing complex dynamic operations to improve performance.

RESULTS

Commercial Directory

It displays a commercial patent data layer for San Carlos county, so users can query POIs information using mobile devices.

WFS was used to generate GeoJSON data to be efficiently displayed on mobile devices using Leaflet library.

AgroMAG

It aims at the creation of data histories from farm plots intended for a particular changing agricultural activity.

It includes plots historical records management in Postgres. The use of database triggers was necessary to make copies of the structure and register new historical values.

Agricultural Geoportal

SDIs could have limitations for private purposes because Geoserver WMS and WFS use are intended for public information management. This application implements data access control scheme using GeoFence as an authentication engine to allow a private access to layers.

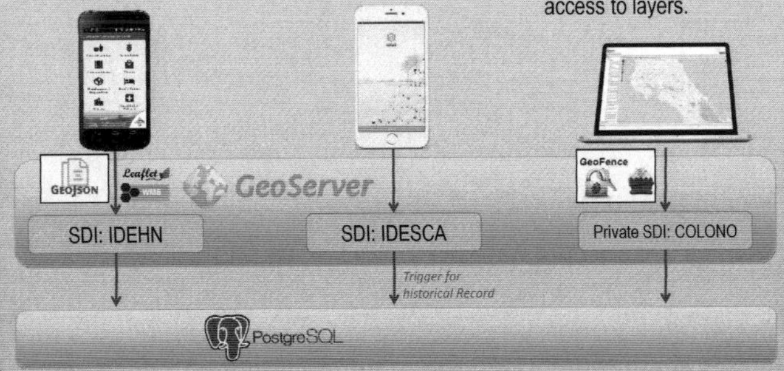

CONCLUSIONS

- The implementation of an SDI for institutions or local governments in Costa Rica requires few resources and basic IT technical knowledge.
- SDI's geoservices allow efficient data access to develop multi-purpose software applications.
- Specific applications could provide non-specialized users better knowledge of the geographic and cartographic areas of their territory.
- For a region with historical limitations, making software using OSGeo's open source tools helped to reach particular objectives at minimum cost in a relatively technical simple way.
- Based on our experience, private and public applications can be implemented using OSGeo's open source tools and miscellaneous libraries with basic programming skills.

REFERENCES

- Arnhardt, C., *et al.* (2007). Sensor based Landslide Early Warning System-SLEWS. Development of a geoservice infrastructure as basis for early warning systems for landslides by integration of real-time sensors. *Geotechnologien science report, 10*, 75-88.
- Kuhn, W. (2005). "Introduction to Spatial Data Infrastructures". Institute for Geoinformatics, University of Münster. Alemania.